SpringerBriefs in Electrical and Computer Engineering

Series Editors

Woon-Seng Gan, School of Electrical and Electronic Engineering, Nanyang Technological University, Singapore, Singapore

C.-C. Jay Kuo, University of Southern California, Los Angeles, CA, USA

Thomas Fang Zheng, Research Institute of Information Technology, Tsinghua University, Beijing, China

Mauro Barni, Department of Information Engineering and Mathematics, University of Siena, Siena, Italy

SpringerBriefs present concise summaries of cutting-edge research and practical applications across a wide spectrum of fields. Featuring compact volumes of 50 to 125 pages, the series covers a range of content from professional to academic. Typical topics might include: timely report of state-of-the art analytical techniques, a bridge between new research results, as published in journal articles, and a contextual literature review, a snapshot of a hot or emerging topic, an in-depth case study or clinical example and a presentation of core concepts that students must understand in order to make independent contributions.

More information about this series at http://www.springer.com/series/10059

Walt Truszkowski · Christopher Rouff ·
Mohammad Akhavannik · Edward Tunstel

Robot Memetics

A Space Exploration Perspective

 Springer

Walt Truszkowski
Goddard Space Flight Center
Hyattsville, MD, USA

Mohammad Akhavannik
Kirkland and Ellis LLP
Washington, DC, USA

Christopher Rouff
Applied Physics Laboratory
Johns Hopkins University
Laurel, MD, USA

Edward Tunstel
United Technologies Research Center
East Hartford, CT, USA

ISSN 2191-8112   ISSN 2191-8120 (electronic)
SpringerBriefs in Electrical and Computer Engineering
ISBN 978-3-030-37951-3   ISBN 978-3-030-37952-0 (eBook)
https://doi.org/10.1007/978-3-030-37952-0

This Springer imprint is published by the registered company Springer Nature Switzerland AG
The registered company address is: Gewerbestrasse 11, 6330 Cham, Switzerland

Acknowledgements

The development of this perspective on robot memetics was started by the late Walt Truzskowski while he was Emeritus at the NASA Goddard Space Flight Center. It was motivated by, and represents, an extension of years of research at NASA on distributed computing, artificial intelligence, intelligent agents, and swarms. A number of people at NASA and other institutions contributed to the motivating research, and we are glad that we are able to continue to build on it. We thank the reviewers for their time spent reading a draft of this publication and the many thoughtful suggestions they provided. Their time is greatly appreciated and their comments have improved this work and given us new thoughts on memetics and intelligent robotics.

Though Walt is no longer with us, we have very fond memories of working with him. We immensely enjoyed the many evenings we spent at his house talking about future space missions, robotics, memetics, artificial intelligence, and many other thought-provoking discussion topics. We will miss him very much.

Contents

Introduction

In the not-too-distant future, the settlement of Mars will commence, eventually involving a joint human–robot partnership coexisting in a hyper society. The Introduction provides a motivation for intelligent robots working in collaboration with humans, forming their own communities and culture based on the goals and the needs of that joint community. The idea of using memetics is introduced as the basis of an intelligent society of robots employing memes to exchange knowledge and form a joint culture with humans. Memes are introduced as a mechanism for information exchange, not only for humans but also for robots as they form a culture and create intelligence based on the needs of their community. An overview of the book describes a path for achieving joint robot–human communities for space exploration and applications on Earth.

Imagine in such a future, these robots can assume the useful and beneficial behavior patterns of other robots and humans to improve and extend their effectiveness and capabilities. Imagine robots that can create and learn behavior bound by societal ethics. Imagine that we have encoded the behavior and intelligence of such robots in a manner offering transparency to robot developers and general human observers alike, thereby facilitating a means for "robopsychology" that goes beyond debugging software to enable explainability and analysis of higher robotic intelligence. Imagine that such transparency was also offered to other robots for that matter. Imagine a degree of human–robot symbiosis permitting faculties ranging from skills to mannerisms to be shared among humans and their collaborative robots toward common objectives (Fig. 1).

This hyper society of humans and robots will grow and flourish for the benefit of all concerned. But, how will all this happen? There may be many ways for achieving this hyper society base on Mars, but answers to that question are being deliberated by the space exploration community at large. The focal discussion herein is about a capability for a highly intelligent robot community envisioned as an integral part of the hyper society. This capability could possibly play a very important and dramatic part in the development and evolution of human–robot relationships.

Fig. 1 Robot expressing a learned mannerism

In the envisioned hybrid community, humans and intelligent robots will need to collaborate to realize mutual benefits as well as to meet exploration and scientific objectives. As such, the concept of memetics will be important to consider. The term *memetics* may be defined as the science that studies the replication, spread, and evolution of memes. In a general sense, the term *memes* may be best understood as referring to pieces of cultural information that pass from one entity to another, but gradually scale into a shared social phenomenon. Memes are a methodical means for exchanging information between individuals and cultures. Essentially, memes are ideas that evolve according to principles similar to those of biological evolution.

All ideas that exist within an individual's mind are examples of memes. In communities, memes have been studied to understand and enhance group learning. Memetic learning works by transmitting these units of cultural ideas or symbols from one mind to another. Although memes spread on a micro-level, their impact is on a macro-level; that is, memes shape the mindsets, behavior, and actions of social groups and change the culture of those groups.

A meme construct can also be used in computer systems as a knowledge base or to form the basis of ideas for robots. This has been a recent subject of interest as a new method for information exchange between computer systems [1–5]. Using memes as a unit of information exchange between robots would facilitate and enable self-adaptation and cooperative behavior necessary for robots to team not only with each other, but with humans as well. We explore how memes can be used in this manner, as opposed to only being used as a means for people to communicate ideas to each other.

This exposition on robot memetics brings a new perspective on the use of the meme construct, providing insights into the utility and conceptual implementation of memes for practical robotics applications, and it provides a new way of thinking about how to realize higher levels of intelligence and learning in robots and robot communities. Ultimately, this is about memes and memetics as elements of collective robotic intelligence that make it possible to achieve this future hybrid community of people and robots working together.

While most scholarly thought about memetics has been focused on its implications for humans, this text speculates on the role that memetics can play in robot communities. The evolution and spread of memes give rise to a culture for a community of robots and humans. Though speculative, the concepts herein are based on proven advanced technology and research on multi-agent systems, artificial intelligence, and robotics. Memes are discussed as constructs for adding intelligence to robots and software "robots". Also considered is the manner in which memes can be adapted and evolved to account for exhibition of intelligent behavior when new and unforeseen situations are encountered, as can occur in many long-term embodiments of intelligence in various environments. With a focus on the emerging culture among robots, a thoughtful perspective is provided on the concept of memetics as applied to the development and evolution of intelligent robots and robot communities and cultures.

Thoughts on robot memetics are offered across a series of chapters conveying fundamental notions and their potential for development and application to real-world human–robot endeavors. As a backdrop and context within which to appreciate the potential impact of robot memetics, a future Mars settlement is adopted in the penultimate chapter, offering a space exploration perspective. The exposition begins with an initial consideration of robots that have been used for space exploration, setting into context the degree of any intelligence they embodied relative to descriptions of automation versus autonomy versus autonomic robots, and positing the potential future implications for memetic robots. Memes and culture are then discussed, that is, how memes have spread throughout history and how the Internet has influenced the spread of a particular notion of memes most recently. To that end, the concepts of Internet memes, intelligent memes, and the use of memes to represent ideas are discussed. Ideas supporting the use of memes for intelligent robotics are then presented, considering previous work, different considerations regarding the use of memes for robots versus memes for humans, how memes can be used to develop a culture for robots, how memes can add intelligence to robots, and the implications of "bad" memes for intelligent robots. The enabling construct of memetic algorithms is then discussed along with supporting ideas including how memes can be observed, how memes can contribute to self-awareness, how memes can be evaluated, and how memes can be formally represented for robots.

The manifestation of presented thoughts and concepts are finally illustrated within a planetary surface settlement scenario on Mars. Such a scenario invites focused consideration of what the characteristics and implications may be for advanced robotic intelligence within a community driven solely by human and robotic intelligence. Multiple descriptive scenarios convey how robot memetics could impact a human–robot community during establishment and evolution of a settlement on Mars. Robot memes and memetics concepts are illustrated within the operational dynamics of a settlement wherein the community is made up of a hybrid culture of robots and humans. The scenarios unfold in ways suggesting how memes can be used to provide intelligence to the robots, how memes can form a basis for learning and representation for ideas, and how memes can be used to change the culture of the hybrid robot and human community to move the group's work in constructive directions while overcoming obstacles so that mission goals are accomplished.

The authors have enjoyed deliberating and writing about memes, robot learning, robot communities, and how robots and humans might interact in the future to achieve new goals. We hope you, the reader, find it as interesting and stimulating in its perspective on the potential of memetics as a facilitator of advanced robotic intelligence and future human–robot collaboration.

References

1. Aunger R (2002) The electric meme: a new theory of how we think. Simon and Schuster
2. Blackmore S (2002) The meme machine. Oxford University Press, USA
3. Hougen DF, Carmer J, Woehrer M (2003) Memetic learning: a novel learning method for multi-robot systems. Robotic Intelligence and Machine Learning Laboratory, School of Computer Science, University of Oklahoma, Norman, OK
4. Silby B (2000) What is a meme? Retrieved from http://www.def-logic.com/articles/what_is_a_meme.html
5. Wilson E, Unruh W (2011) The art of memetics. lulu.com

Chapter 1
Spectrum of Robots

1.1 Introduction

A number of working definitions of "robot" are used today. Herein, we adopt the following definition: A *robot* is an electromechanical system that exists in a physical world; using electronic circuitry, embedded computers, sensors, and actuators, it senses itself and its environment and is self-controlled or human-controlled to physically act on the environment to achieve a goal. A related term is *agent*, which in the context of computer science refers to a software agent, that is, a computer program that makes decisions on the appropriate action(s) to take on behalf of a user. Today, research on autonomous robots and autonomous agents shares a common focus on cognitive functions and organizational activities including inter-agent communication, negotiation, coordination, conflict resolution, and social behavior. Agents are more typically disembodied rather than physically embodied in moving hardware, as is the case for robots.

We consider three categories of robots:

- *Automated*—robots for which routine manual processes are replaced with hardware and/or software processes that follow a step-by-step sequence that may involve human participation.
- *Autonomous*—robots for which human involvement is eliminated during task execution but necessary for robot management (fault management, sustenance, maintenance, etc.). Such robots make independent decisions to achieve goals that emulate human processes and are considered capable of performing behaviors or tasks through independent action (*i.e.*, without a necessity for human involvement or intervention).
- *Autonomic*—robots that are autonomous as well as capable of self-management, where self-management involves the properties of self-awareness, self-situational (context) awareness, self-monitoring, and self-adjustment. Such robots may take on human-like qualities for cognition and reasoning.

W. Truszkowski et al., *Robot Memetics*,
SpringerBriefs in Electrical and Computer Engineering,
https://doi.org/10.1007/978-3-030-37952-0_1

This "triple-A" categorization spans a spectrum of robotic intelligence with embodiment ranging from automated robots through autonomic humanoid robots.

A range of robots have been used for space exploration. Perhaps the most recently familiar are the planetary robotic rovers Spirit, Opportunity, and Curiosity that roamed the surface of Mars during missions lasting multiple years since the early 2000s (Fig. 1.1), and the Yutu rovers landed on Earth's moon. Others include the spacecraft that transported them to their destinations or spacecraft that have traveled to other planets, asteroids, or to outer reaches of the solar system. Even spacecraft in Earth orbit, such as the Hubble Space Telescope, could be classified as robots or unmanned spacecraft that point telescopes and other instruments at distant objects or areas of interest to photograph or remotely sense celestial bodies and phenomena.

Space exploration has thus far employed unmanned spacecraft and robotic vehicles of various types. These include orbiter and fly-by spacecraft, planetary landers, and planetary rovers (Fig. 1.2). Their use has resulted in a growing body of knowledge about our solar system, its sun, and various small and large bodies. Many of the platforms employed for robotic space exploration are designed for in situ operations. They build on the successes of earlier reconnaissance spacecraft that performed fly-by and orbiting missions throughout the solar system. Subsequent missions aim to acquire successively more scientific knowledge by delivering instrumented platforms into planetary atmospheres and/or onto the surfaces of planets, moons, and small bodies, such as asteroids and comets. Future missions are planned to go a step

Fig. 1.1 Prototypes of Mars rovers for NASA missions (left to right): Mars Exploration Rover (Spirit/Opportunity), Mars Pathfinder rover (Sojourner), and Mars Science Laboratory rover (Curiosity). Courtesy: NASA/JPL-Caltech

Fig. 1.2 Images and depictions of NASA unmanned and robotic spacecraft: Top—Dawn space-craft, Curiosity Mars rover, New Horizons spacecraft; Bottom—Voyager spacecraft, Surveyor lunar lander. Courtesy: NASA, JPL-Caltech

further to acquire and analyze physical samples of planetary surface and subsurface material in situ, and some seek to return collected samples to Earth for analysis in comprehensive laboratories. We focus here on the spectrum of mobile robots that have been used to conduct planetary surface exploration missions.

Within the context of planetary surface exploration and science missions, planetary robotic rovers that are designed to be teleoperated and those designed with any amount of embedded autonomy are considered robots. Generally speaking, a rover refers to a mobile robot that is capable of roving about on natural terrain surfaces. Planetary rovers take on a variety of configurations and purposes typically governed by their specific missions and target environments. Thus far, rovers have been the most popular solution for planetary surface mobility. While all rovers that have been operated on space flight missions to date have employed wheels for locomotion, tracked and legged locomotion systems are viable alternatives for some applications and can offer certain advantages in very rough or steep terrain.

Space agencies around the world employ robotic manipulators and mobile robotic vehicles (rovers), instrumented with a variety of sensors and tools, as surrogate explorers on remote planetary surfaces, on orbit, or as assistants to astronauts. The utility of space robots is a function of their ability to move about and explore intelligently without frequent contact with or strong reliance on human mission operators. As such, robotic autonomy and success of robotic space missions is closely related to space robot capabilities of sensing and perceiving their surrounding unstructured or uncharted environment, and their abilities to perform safe, reliable, and rational actions to accomplish mission objectives. The variety of robots used to perform space missions to date (Fig. 1.3) include the Soviet Union's Lunokhod rovers teleoperated from Earth in the 1970s, the numerous NASA Space Shuttle flights that consistently operated Canada's Shuttle Remote Manipulator System (Canadarm) since the early 1980's, Germany's ROTEX technology experiment in 1993, Japan's Experimental Test Satellite VII in 1998, America's Orbital Express on-orbit satellite servicing experiment, America's Mars Pathfinder rover technology experiment in 1997, twin Mars Exploration Rovers during 2004–2019, Mars Phoenix polar lander in 2008, Mars Science Laboratory's Curiosity rover since 2011, China's lunar rovers in 2013 and 2019, and a variety of robots operating on the exterior and interior of the International Space Station including robotic manipulators, free-flying inspection robots, and NASA's Robonaut-2. With successful uses of space robots in missions over the past several decades, telerobotics and robot autonomy technology have been given a proving ground for demonstrating the utility and practicality of robots for performing tasks in the real world. Each successful robotic mission raises the bar for subsequent missions and increased space robot intelligence becomes more and more essential for accomplishing missions of greater demand and complexity.

Fig. 1.3 Images and depictions of space and planetary robots used on various missions: Top—Lunokhod lunar rover, Space Shuttle Atlantis Canadarm with astronaut attached, AERCam Sprint free-flying robot outside the cargo bay of the Space Shuttle Columbia, Robonaut R2 on the International Space Station; Bottom—ETS-VII spacecraft and Orbital Express spacecraft performing robotic satellite servicing operations, Phoenix Mars lander, and Chang'e-3 Yutu lunar rover. Courtesy: NASA, NASDA/JAXA, CNSA

1.2   Primeval/Nascent Planetary Rovers

Both teleoperated and semi-autonomous robots have been employed thus far for planetary surface missions to Earth's moon and Mars. Planetary surface exploration is one of the few applications for which autonomous robot technology has met with success. On planetary surfaces, robots have thus far been used for scientific exploration and will later perform service, construction, and other utility tasks associated with their use on precursor human missions, performing work and building infrastructure for eventual human habitats. As it becomes feasible for humans to have extended stays on planetary surfaces, robots will continue to be useful as tools and partners in human settlements, and will ultimately exist as members of a society of humans and robots. Relative to this future, planetary rovers to date can be considered primeval or nascent in a "techno-evolutionary" sense. Considering the triple-A robot categorization established earlier, most planetary rovers to date hold membership in the automated category with a few exceptions holding membership in the autonomous category. Their assessment as primeval is particularly appropriate relative to future autonomic humanoid robots. Such neo-humanoid robots will have levels of artificial intelligence and cognition sufficient for participating with other robots and humans as legitimate members of the future human–robot society. *To realize very intelligent humanoid robots of the future, much more has to be done to advance robotics technology in the coming decades.* For the moment, let us consider progress to date and where technology stands today along the spectrum of planetary robots.

Among the earliest exploratory development of mobile robotic intelligence for possible planetary surface exploration was work done in the 1960s on mobile automata as adaptive machines [1, 3]. Engineers at the Johns Hopkins University Applied Physics Laboratory developed two mobile robots (Fig. 1.4), each with primitive intelligence aimed at the objective of independent operation for relatively long

Fig. 1.4   Johns Hopkins University Applied Physics Laboratory mobile automata: Ferdinand, Mod 1 (left); Beast, Mod II (right) (© The Johns Hopkins University Applied Physics Laboratory.)

periods of time in unmodified environments and with provision for two survival mechanisms: the ability to replenish its energy supply whenever necessary and to cope with obstacles in its path. This early research was ultimately aimed at enabling capabilities for future rovers that might traverse the surface of Mars and explore the depths of Earth's oceans. The early stages of the research addressed the problem within the more manageable experimental scope of indoor navigation in the environment of an office building, such that the prototype mobile automata were designed to safely roam hallways of an office building and to plug into electrical wall outlets to recharge when necessary. On these robots, built using analog circuitry and computing, a suite of tactile sensors enabled the primitive sensor-based intelligence for obstacle avoidance (after detection of momentary obstacle contact), stairway avoidance (upon detecting immediate loss of floor support), and electrical outlet search and docking behaviors (by feeling along walls using a retractable arm in a manner akin to a blind person's use of a guide cane or when using fingers to discern raised features as when reading Braille characters). Early experiments with the first prototype, named Ferdinand (Automaton Model I, or Mod I), exhibited survivability for tens of hours in a closed hallway environment while remaining energized over the course of tens of self-charging cycles, roaming hallways at 7–15 cm/s. The second prototype, the Hopkins Beast (or Mod II), roamed twice faster and was upgraded with sensors including non-contact acoustic and optical/video guidance, as well as brain-inspired neuronal analog electronic devices that would facilitate guidance and control based on continuous sensing rather than discrete tactile sensing alone. These early robots were encoded with multiple behavior patterns comprising a basic behavioral system intended, in the long run, to foster sophisticated learning. The research vision had its focus on future mobile automata that would be sufficiently intelligent to explore remote or hostile environments without direct human control or supervision. These developments represent some of the earliest instantiations of embodied artificial intelligence following seminal mobile robots such as the Stanford Cart. During the same period, the space program in Soviet Russia was developing rovers for exploration of Earth's moon and for evaluation of terrain properties on Mars, spawning a lineage of planetary rovers to follow over the course of subsequent decades.

The Russian Lunokhod (i.e., moonwalker) rovers were primarily designed to support the Soviet program of human missions to the moon which were ultimately canceled. The rovers were later used as teleoperated robots for lunar surface exploration and lunar surface image acquisition following the successful NASA Apollo human missions to the moon. Lunokhod 1 (the first teleoperated planetary robotic rover, delivered to the lunar surface in late 1970) ran during the lunar day, stopping occasionally to recharge its batteries via its solar panels. At night the rover hibernated until the next sunrise, heated by its radioisotope heater unit. Lunokhod 2 was the second and more advanced of the Soviet Union's two unmanned lunar rovers delivered to the lunar surface in early 1973. Images acquired by its onboard cameras were used by a five-man team of controllers on Earth (driver, commander, navigator, radio antenna operator, and the flight engineer) who sent driving commands to the rover in real time. The Lunokhod rovers were not endowed with cognitive or autonomous capabilities.

Fig. 1.5 PrOP-M (Russian
acronym for Device
Evaluation Terrain—Mars)
rover

During the same time frame, in 1971, the Soviet Union sent two spacecraft landers
to planet Mars as missions called Mars 2 and Mars 3 (Fig. 1.5). Both of the Mars 2
and Mars 3 landers carried a small Mars rover (PrOP-M) on board (mass of 4 kg,
~20 cm in largest dimension), which was designed to move across the surface while
connected to the lander via a 15-m power and communications tether. The mobility
system, comprised of ski-like mechanisms, would enable locomotion in a sort of
slow ski-walking manner (at up to 1 m/h). To enable measurement and evaluation of
terrain properties near its lander, the rovers had both a densitometer and a dynamic
penetrometer for testing the density and bearing strength of the soil. The intelligence
embodied in the PrOP-M rovers was limited to tactile sensor-based autonomous
obstacle detection (by contact) and avoidance. These rovers would have used these
features to explore the area within 15 m of the lander, however, neither PrOP-M
rover got the chance. The Mars 2 lander unfortunately crashed during descent to the
Martian surface, and the Mars 3 lander suffered permanent loss of communications
within one minute of safely landing on Mars.

A couple of decades passed before another robotic rover was sent to explore a
planetary surface. In 1997, the NASA Mars Pathfinder mission landed the rover,
Sojourner, a technology demonstration of a way to deliver an instrumented lander
and a free-ranging robotic rover to the surface of Mars. With a concept of operations
similar to that intended for the PrOP-M rovers, Sojourner was designed to explore
the Martian surface local to the Mars Pathfinder lander, however, untethered and
within 500 m, a range limited by the lander-rover communications link. The lander's
communications system was the means through which the Sojourner rover received
command sequences from Earth and by which its telemetry and acquired science data
were sent to Earth. This rover incorporated modern digital computer technology in
the form of a microprocessor running embedded software executing a behavior-based
control system, an artificial intelligence paradigm for mobile robots. Sojourner's level
of intelligence enabled safe, autonomous waypoint navigation (at 1 cm/s) in rough

and rocky terrain in the vicinity of the Mars Pathfinder lander using non-contact-based obstacle detection and avoidance facilitated by optical sensors (cameras and lasers). This capability enabled the rover to navigate on its own to locations of scientific interest near the lander. Such locations are designated by human operators in uplinked command sequences for scheduled execution by the rover.

Several years later in the summer of 2003, NASA revisited the Mars surface with two separate launches of the twin Mars Exploration Rovers, named Spirit and Opportunity, delivering them to landing sites on nearly opposite sides of the planet in, respectively, early and late January 2004.[1] These rovers were designed with more sophisticated autonomy and for longer duration missions than Sojourner. And unlike Sojourner, their missions were operated independent of their landers as communications links to Earth—they carried all communications equipment on board enabling them to communicate directly to Earth and indirectly via relay satellites already in orbit around Mars. Spirit and Opportunity were equipped with a suite of science instruments enabling them to be operated effectively as surrogate, remote field geologists for an Earth-based science team exploring the landing sites for signs of past water activity on the Martian surface. These rovers incorporated embedded microprocessor-based computer technology running a real-time operating system with functionality ranging from low-level fault detection to higher-level robotic autonomy. Autonomous capabilities for Spirit and Opportunity included autonomous waypoint navigation, autonomous manipulator-mounted science instrument placement, odometry and navigational target tracking based on computer vision algorithms, and global path planning based on artificial intelligence algorithms. Their level of intelligence enabled safe mobility, navigation, and robotic arm operations in terrains of varied complexity over kilometers within the regions of their landing sites. Like Sojourner, Spirit and Opportunity maneuvered through their environments using non-contact-based obstacle detection and avoidance facilitated by optical sensors (in this case, multiple pairs of cameras for 3D stereovision). This capability enabled the Mars Exploration Rovers to navigate on their own to human-designated locations and areas of scientific interest throughout their landing regions, including over hills on the horizon and into small and large craters. Uplinked command sequences for scheduled execution by each rover comprised a day's worth or more of commands to be executed autonomously between opportunities for communications with humans on Earth.

The end of Spirit's mission was announced around the same time as NASA's scheduled launch of a next-generation rover named Curiosity on its Mars Science Laboratory mission. The Curiosity rover is larger in size than the Spirit and Opportunity rovers, powered by a radioisotope power system rather than a solar panel, and equipped with a more advanced suite of science instruments, tools, and an onboard laboratory. The rover's science equipment facilitates data and soil/rock sample acquisition and associated onboard analysis (of chemistry in acquired samples) permitting

---

[1] The last contact with rover Spirit was on March 22, 2010 and the last contact with rover Opportunity was on June 10, 2018. The end of the mission was declared on February 13, 2019.

Earth-bound scientists to assess the habitability of Mars, that is, whether the Martian environment was or is able to support microbial life. In its role as a surrogate robotic field geologist for an Earth-based science team, the Curiosity rover uses computing, communications, and operations architectures similar to those employed to explore Mars using Spirit and Opportunity. Curiosity inherits its autonomous capabilities from Spirit and Opportunity, although visual target tracking and global path planning software that provided additional autonomy was first uploaded to those rovers after several years into their missions in order to test and prove the capabilities on Mars for use by future rovers such as Curiosity. Since being active on the surface of Mars, Curiosity has been endowed (via newly uploaded software) with additional autonomy enabling opportunistic science, that is, artificial intelligence for autonomous selection and measurement of terrain features visually classified as scientifically interesting. Just as Spirit and Opportunity were designed for longer duration missions than Sojourner, Curiosity was designed for a mission of even longer duration involving longer distance driving to visit a greater diversity of sites, and more complex robotic arm interaction with terrain while acquiring samples by scooping soil or drilling into rocks. As of this writing, the Mars Science Laboratory mission is still underway.

In 2013, the Chinese Lunar Exploration Program landed its Chang'e-3 lunar mission lander on the moon, also delivering China's first lunar rover, Yutu, to the surface. This fulfilled the program objective to demonstrate technologies for China's first soft-landing and for roving exploration on the Moon. Early in the mission, Yutu executed short distance drives near its lander in order to photograph the lander from several different angles, later traversing tens of meters across the surface. After about one month on the lunar surface, Yutu was unfortunately announced immobile due to rover hardware operational difficulties. It was reported that Yutu's mobility was lost on the rover's second day on the moon. Like the NASA Mars rovers, Yutu's robotic intelligence included computer vision-based obstacle detection and avoidance and artificial intelligence algorithms for path planning.

Several years later, China launched its Chang'e-4 mission in early December 2018, which landed on the far side of the moon in early January 2019 equipped with the second generation of its lunar rover, Yutu-2, modeled after its predecessor Yutu. Designed to last for three lunar days (about three Earth months), Yutu-2 supported the Chang'e-4 mission objective to better characterize the environment of the lunar far side and understand how it differs from the more familiar near side. Surviving with operational status for longer than its design lifetime, Yutu-2 traversed several hundred meters on the lunar far side surface as of 8 months since the landing. By then, the Indian Space Research Organization (ISRO) launched its Chandrayaan 2 spacecraft in early September 2019 carrying India's first Vikram lander and planetary rover, Pragyan. The ISRO Vikram lander targeted to land near the lunar South Pole, and the Pragyan rover were designed to last one lunar day on the surface. The Vikram lander unfortunately crash-landed and Pragyan was never deployed. The solar-powered Pragyan rover was designed to travel up to 500 m and would have used algorithms driven by structured light sensing for 3D terrain map generation and mobility hazard detection to facilitate rover path planning.

The most recent of these planetary rovers feature elements of robotic intelligence representative of the state of the art. Currently, the state of practice for planetary robotics lags that of Earth-based robotics. That is, advanced component technologies and capabilities that are close to the cutting edge are more readily employed for current robotics applications on Earth than for applications targeted for other planetary surfaces. An example is the mature navigation technology that is enabling higher-speed mobility of autonomous road vehicles including personal automobiles and commercial ground transportation vehicles. This is due in part to the risk-averse nature of space mission planning and execution (e.g., associated with preferences to use electronic and other hardware that has already proven reliable in space environments rather than the latest and greatest technology yet to be proven on a space mission). It is also due in part to pragmatic issues such as the lack of availability of high-end computing and avionics for planetary robots that can actually survive the radiation, thermal, and otherwise harsh environments of planetary destinations.

Today's state of the art for Earth-based robotics and current trends for continued research and development serve as a foundation for future planetary surface robots that will exhibit higher-level intelligence. That foundation includes autonomous vehicles that can navigate safely at high speeds in certain structured and unstructured environments, robots that can cooperate to perform joint tasks, robots that collaborate with humans in close proximity and in direct physical contact, and increasingly capable humanoid robots that are effective in human-centric environments using tools designed for human use under supervised autonomous control of human operators. Also representative of current state of the art for Earth-based robotics is legged robots capable of highly stable dynamic motion control, increasingly dexterous arm- and hand-based manipulation (facilitating object handling and assembly tasks for varieties of service robots), robotic surgery systems, and emerging capacities for robot learning and skill acquisition from data, from experience, and via demonstration or observation (by or of humans or other robots). Substantial capability exists today as well for automated planning and reasoning systems building toward improved means for realizing cognitive robots. Ultimately, a cognitive robot would have the ability to mimic cognitive mechanisms of a brain and would thus be capable of inference, perception, and learning [6]. Such mechanisms, developed sufficiently for a range of applications, would enable cognitive robots to function reliably in unstructured environments by equipping them with abilities to recognize, learn, reason, decide, and communicate.

A number of these features are embodied in present-day humanoid robot prototypes and undergoing technology advancement. An exemplar prototype with the distinction of being developed with future space and planetary missions in mind is the NASA humanoid robot R5, also known as Valkyrie (Fig. 1.6), the latest generation in the Robonaut lineage. Current R5 prototypes were designed for robust operability in human-engineered environments that are damaged or otherwise degraded. The multifaceted utility of such humanoids for future missions includes building habitats on the Martian surface, mining resources for a Mars settlement, doing emergency repair, and completing disaster-relief tasks. At times they would work on their own, with other robots, alongside humans, or in close cooperation with humans. For now, a

Fig. 1.6 NASA R5 (Valkyrie) humanoid space robot prototypes. Image credit: Kris Verdeyen/NASA

goal of continued technology development is to increase the robotic intelligence and autonomy for dexterous humanoid robots to levels sufficient for performing complex tasks, such as those mentioned above, on future missions on Mars and elsewhere.

1.3 The Future

Future planetary rovers and humanoid robots will leverage a greater share of the Earth-based robotics state of the art as technologies mature and exhibit greater robustness and reliability, and as further experience is gained with the deployment of surface robots on other planets, thus reducing the associated risk posture. As noted earlier, and worth reiterating, *much more has to be done to advance robotics technology to a point where very intelligent robots of the future can be realized.* Today's emerging technologies and relevant advancements will play a part, including computing hardware and software, artificial intelligence, machine learning algorithms, autonomy, and autonomic systems. Such technologies will equip future robots with the fabric for accommodating memetic algorithms and they will furthermore be augmented with powerful capabilities enabled by robot memetics.

As part of the human expedition beyond Earth, a future society of humans and robots will include a population of very intelligent humanoid robots of the autonomic category that will serve as collaborators with humans in the establishment of a working settlement on Mars. For such robots to be considered very intelligent, they must have some form of developmental and evolvable, artificial "brain". That brain must have capacities for continual robot learning and adaptation, beyond the level of intelligence the robots have when first delivered to the surface of Mars. Their brain must have some semblance of the complementary functionality attributed to both halves of the human brain. This will allow for creativity and some degree of individuality that emerges from their developmental and evolutionary capacities. So technologically equipped, the very intelligent humanoid robot population within the human–robot society on Mars should be able to manage language, individual mental existences, individuality, and collectively foster the development of robot culture. As such, they should be sufficiently socially sophisticated to interact and exchange information and capabilities with other robots and humans. With such sophistication and capacity, very intelligent robots could even have psychopathologies, life histories, families, and more.

As very intelligent robots adapt and learn via their individual experiences at the Mars settlement, while also observing and interacting with humans and other robots, they will further develop through the technology of robot memetics. The resulting dynamic exchange of information and capabilities among robot members of the settlement will ultimately establish the settlement's human–robot culture. Humanoid robots may assume the highest levels of robot intelligence and capability, particularly regarding capabilities learned and memes acquired from humans due to their kinematic similarity in physical form. This would garner potential for humanoid robots to become the leaders of other robots in such a settlement, especially if they are equipped with greater embedded computational resources and thus greater capacity enabled by combined intelligence and physical versatility. Social hierarchies among robots could then be expected to manifest wherein humanoid robots serve as foremen or managers of activities performed by other robots and as principal assistants to humans in broad areas of human–robot endeavor. All of this and more should be possible with memetic technology as an enabler.

1.4   Summary

For future missions on planetary surfaces, networked teams of robots will work cooperatively to prepare sites for human habitats at permanent planetary settlements. Robots will work with automated machines and instruments as an integrated system that constructs, services, monitors, and repairs planetary outpost facilities. Such tasks may be shared with astronauts as planetary surface settlers. Robotic systems will also help to automate prospecting, mining, and processing of raw materials to help sustain human presence by minimizing dependence on Earth support. Automation and robotic technologies to be developed toward this end may be used in similar ways to facilitate eventual human exploration and settlement of other locations on the same and other planetary bodies. As these future settlements will be designed for human presence and use, the variety of robot types will include advanced humanoid robots. Humanoid robots will be able to use the same tools as those designed for human use and will operate effectively in facilities of human-centric design without requiring modification or tailoring to accommodate particular robot characteristics. Planetary surface robotic systems will advance to progressively higher levels of autonomous capability enabled by new and improved sensing solutions, algorithms, and computing enabling higher levels of robot automated reasoning, cognition, and learning. For maximum effect in communicating with humans, robots will be capable of interacting via language, gestures, and brain–machine interfaces.

Relative to this future, the planetary robots of today can be considered "primeval", ancient robot ancestors of the brand of robots characterized in later chapters herein. Beyond the modest onboard intelligence and autonomy realized today, future robots and planetary variants, including rovers and humanoids, will be more *self-aware*. That is, robots will possess information about their internal state and will have sufficient knowledge of their environment to determine how *they* are perceived by other robots and agents, or by humans. A robot with such self-awareness along with the means to monitor and make adjustments to itself will have the capacity to self-configure, self-heal, self-optimize, and protect itself from conditions that would reduce its intended level of performance. In short, with such characteristics, robots will be capable of self-management, and thus autonomic [2, 4, 5].

As humans and the necessary transport vehicles are prepared for human Mars missions, utility rovers and humanoid robots will conduct precursor missions on Mars to build and prepare outposts with infrastructure and habitats to support eventual human presence. When humans arrive, the robots will work for and with humans throughout the outpost and its surroundings. These robots will be smarter and more capable descendants of rovers discussed above and nascent humanoid robots, such as the lineage of NASA Robonaut systems. They will be autonomic and capable of learning. They will be adaptive to their environment and to new situations, and their behaviors will adapt in response to behaviors and mannerisms of the robots and humans with whom they interact. They will also have the means to mutually share and transfer knowledge through robot memetics.

A foundation for this future has been lain by nascent mobile automata as adaptive machines, the short lineage of planetary rovers with modest intelligence enabling safe autonomous mobility in remote and hostile environments, and the various present-day humanoid robot prototypes being developed as partners to humans. Consider the potential outcomes that could be realized if the "primeval" robots developed over the past fifty years were uploaded with software to embed autonomic properties, higher intelligence, and greater autonomy. Further consider the realm of the possible if such robots populated a common environment with humans and had the sophistication to share information and capabilities, thus learning from each other and from humans. It would not be hard to further conceive the evolution of a human–robot society with an established culture. Such a future is opened up to the imagination in subsequent chapters.

References

1. Carlton AG (1962) The mobile automaton. APL Tech Dig 2(1)
2. Kephart JO, Chess DM (2003) The vision of autonomic computing. Computer 36:41–52
3. Semmel RD (2003) An overview of information processing and management at APL. Johns Hopkins APL Tech Dig 24(1)
4. Truszkowski W, Hinchey M, Rash J, Rouff C (2006) Autonomous and autonomic systems: a paradigm for future space exploration missions. IEEE Trans Syst Man Cybern Part C 36(3):279–291
5. Truszkowski W, Hallock H, Rouff C, Karlin J, Rash J, Hinchey M, Sterritt R (2009) Autonomous and autonomic systems: with applications to NASA intelligent spacecraft operations and exploration systems. Springer
6. Wang Y (2015) Cognitive learning methodologies for brain-inspired cognitive robotics. Int J Cogn Inform Nat Intell 9(2):37–54

Chapter 2
Memes, Culture, the Internet, and Intelligence

2.1 Introduction

Since Richard Dawkins first coined the term "meme" in 1989 and defined it as a "unit of cultural transmission," memes have been a focus of academic study to further understand the culture. The advent of the Internet and Internet culture has sharply changed how memes are shared and culture is shaped. Although Dawkins' definition is conceptual, memes can be formally modeled in a way that will enable them to be manipulated by non-humans. Further, memes can have encoded intelligence which will transform them into dynamic forms of information. We introduce this concept as an "intelligent meme." Creative adaptations of intelligent memes are discussed in this chapter.

2.2 Memes

Richard Dawkins first defined memes as a unit of cultural transmission [6]. The word meme derives from the Greek mimema, meaning "something to be imitated." A meme can be thought of as an idea or behavior that can be imitated and spreads from one entity to another within a culture. Examples of memes are catch phrases, musical themes, scientific ideas, physical actions, and sayings. Memes are used to exchange ideas and information between individuals and cultures. For example, replication and transmission of memes within modern human culture may take place by verbal and written communication, TV, radio, and the Internet.

All ideas that exist within an individual's mind are examples of memes. A meme acts as a unit for carrying these ideas, symbols, and practices that can be transmitted from one intelligence to another through various mechanisms including speech, gestures, rituals, and other imitative phenomena. Memes act as replicators that transmit

W. Truszkowski et al., *Robot Memetics*,
SpringerBriefs in Electrical and Computer Engineering,
https://doi.org/10.1007/978-3-030-37952-0_2

the ideas. As memes are transmitted, they undergo variation, competition, selection, and retention [14]. Only memes that are well-suited to their sociocultural environments are successful and spread, while others become extinct. Memes reflect deep societal and cultural artifacts. Memes can be thought of as pieces of cultural information that pass along from person to person, but gradually scale into a social phenomenon [14].

Memes can also be thought of as cultural analogs of biological genes in that they can be self-replicating, mutate, and respond to selected pressures with a mimicked theme [6]. Memes can evolve according to the same principles as biological evolution. One key difference between memes and genes is that genes can only be transmitted from parents to offspring while memes can be transmitted between any two individuals and can therefore replicate and mutate at a more rapid pace. Memes that are good at replicating leave more copies of themselves in people's minds and therefore expand their influence. Memes also are able to propagate within a society at a much faster rate than genes are able to spread in a community. That is because memes are not restricted in the number of copies they can make or by the time it takes for a new generation of offspring to be created. Due to the differences between memes and genes, it is expected that memetic evolution can occur orders of magnitude faster than genetic evolution.

Since memes are learnable, they have the capability of being taught and shared among members in a community. Memes benefit from their similarity to genes [10]. As previously discussed, memes are also under constant selection pressures like genes. Indeed, they can mutate and combine with others in an evolutionary fashion like genes. Memes are also in constant competition to be absorbed and evolved from the collection of memes in the community. Memes are competing to be learned and those that are better at reproducing achieve their intended behavior and influence, while those memes that are not able to reproduce die. Since memes represent ideas, they can be modified and updated over time to represent the latest cultural trends and newest knowledge that will make them more useful.

Dawkins describes three elements for successfully spreading a meme in an environment: longevity, fecundity, and copy fidelity. Longevity is a concept that describes how memes can be stored over time. Fecundity describes the capability to make new copies or offspring. Fidelity describes the accuracy of a reproduced copy. Memes have a high rate of success when these three characteristics are present.

A famous pre-Internet meme is the "Kilroy Was Here" meme [14, 16] and is a good example of how memes can spread and change over time. It is thought that the meme was started during World War II by a ship inspector named James Kilroy. He would inspect rivets in ships and would write "Kilroy Was Here" inside the ship. Later sailors would find the inscriptions, and not knowing who put them there, started attributing them to a "Super GI" (GI as generically referring to an American soldier) and started inscribing it in other places and it spread with variations to the United Kingdom and Australia during the war. Along the way, it is reported that it was changed to include the cartoon of a person with a long nose looking over a wall (Fig. 2.1).

Fig. 2.1 A depiction of the "Kilroy Was Here" meme on the WWII Memorial in Washington, DC (https://upload.wikimedia.org/wikipedia/commons/9/99/Kilroy_Was_Here_-_Washington_DC_WWII_Memorial.jpg (creative commons license image, author Luis Rubio (http://www.flickr.com/people/44801240@N00.))

As the meme caught on, instead of the meme spreading from writings inside of ships or freight, the meme spread over the radio, through speeches, newspapers, magazines, word of mouth, and other types of communication. People would spread the meme by writing it on walls, fences, or other places where people could see it, and who would then also spread it in a similar fashion. It has now spread across a huge portion of the general population, endured for decades and become part of the western culture, and it is likely to endure for at least a century if not more, all starting from an inspector writing something on a wall to show he had inspected some military equipment.

Leonard Adleman has expanded on the idea of memes and genes to include "cenes" and "prenes" [2]. Adleman defines memes as what is stored in a person's brain, genes as what is stored in a cell, and cenes as a gene or a meme that is stored in a computer. A prene represents the storage of a gene, meme, or cene across mediums. An example of a prene being stored across multiple mediums might be a meme stored in one's mind as well as on a computer, or a DNA sequence stored in a cell or its sequence stored in a computer. Genes, memes, cenes, and prenes all have Darwinian replication and evolutionary characteristics and struggle against extinction, moving between storage media to remain "alive". When a prene is no longer stored in any medium, it becomes extinct. With the proliferation and replication of data on the Internet, making a prene extinct becomes difficult.

Adleman gives an example of a gene moving into a computer that keeps it from being extinct, and has the potential to move back into people in the future. Adleman describes how smallpox, which has been eradicated, is stored in two labs for scientific purposes [1]. The smallpox DNA has now been sequenced and stored in a computer (and can be found on the Internet). Recently a new area of science called synthetic biology [3] has developed technology that can synthetically produce DNA sequences based on a sequence stored in a computer. The DNA sequence itself is not alive, but Craig Venter, a synthetic biology researcher, has claimed to have produced a synthetic cell by introducing synthetic DNA into a host cell [15]. Dr. Venter claimed by doing this that he has created "the first self-replicating species we've had on the planet whose parent is a computer." So, even though smallpox now only exists as a DNA sequence in a computer, it still exists as a cene and is not extinct. It has survived by replicating itself through computers, and if someone were to produce a smallpox synthetic DNA and put it in a human host, it could live again in humans. In the future, we may need to develop digital vaccines to find and eradicate harmful DNA sequences from computers to keep them from re-entering the human race.

In this book we do not make the distinction between memes, cenes, and prenes. We refer to them, whether stored in someone's mind or on a computer, as a meme.

2.3   Complex Memes

A single meme expresses an individual idea. A complex meme is generated when multiple memes are combined to form a new meme. Complex memes can be generated by combining individual memes in different ways. One way in which a complex meme can be generated is by sequencing multiple memes together. A complex meme that is generated by sequencing multiple memes together will perform or convey a series of memes consecutively in order to perform a new function or convey a more complex idea. Alternatively, a complex meme can be generated by superposition such that a new meme is formed having the combined effect of individual constituent memes. The superposition could involve equally full influence from all constituent memes or weighted partial influence from constituent memes. Furthermore, a complex meme could be formed having partial influences from constituent memes that are variable and adaptive to conditions or situations. In such combinations, a complex meme will combine components from different memes to form a new meme to perform a new function.

Dawkins notes that certain groups of memes are co-adaptive and tend to be replicated together so that they can strengthen each other. These groups of co-adaptive memes are called "co-adaptive meme complexes" or "memeplexes" [6]. A memeplex is analogous to a complex meme since the individual memes that make up the complex meme are often replicated together. For example, a complex meme could be an algorithm to perform a multi-step process for identifying and collecting certain materials. The individual memes that make up this algorithm (i.e., the memeplex) would be strengthened and replicated together since they are dependent on each other to complete the algorithm. Another example of a memeplex would be a group of memes that are each useful to achieve a common objective. For instance, a group of individual complementary memes necessary for responding to inclement weather, such as safety and pathing protocols, would be replicated and strengthened by each other because they are co-adaptive and increase the likelihood of one another's success.

2.4 Memes and Culture

The generation and consumption of memes in a community is a reflection of that community's culture. According to the American Heritage Dictionary [4], "Culture" may be defined as the "totality of socially transmitted behavior patterns, arts, beliefs, institutions, and all other products of human work and thought." Cultural products are embodied in thought, speech, action, and artifacts, depending upon an entity's capacity for learning and transmitting knowledge to succeeding generations through the use of tools, language, and systems of abstract thought. Culture is used to describe a community's viewpoints, customs, expressions, rules, and other social norms. Additionally, the way members of a group express themselves physically and communicate with each other through verbal and nonverbal forms is an expression of their culture. Cultural practices are often maintained as traditions across multiple generations and long periods of time. Culture is also used to describe practices within a subgroup of a society.

A society's culture is shaped and guided by its ethical values. A society's ethical values also shape the creation and spread of memes. Members of a society make decisions based on central ethical values that are shared by their peers. The ethical principles that exist in a society therefore also affect the probability of success of a meme's propagation because ethical values serve as rules according to which members decide how to act. The memes that encapsulate a society's ethical values are highly valued and deeply rooted in the minds of members of society. Indeed, memes that go against commonly accepted ethical principles will be suppressed by the community.

Members of a community will typically seek to adopt memes that promote their self-preservation. However, some memes may promote an individual's self-preservation while at the same time harm a community's moral values. Ethical memes act to regulate the propagation of memes that act against the collective goals of a community even though they may be beneficial to an individual. For example, stealing may help an individual but harm the community as a whole.

If a culture does not promote certain principles, such as individuals taking things which do not belong to them, then memes related to thieving are not likely to be adopted by members of that community. Memes that act to prevent the spread of thieving memes may also be propagated. This phenomenon can be commonly witnessed in our society. For example, people look down upon thieves, even if they are acting in self-preservation. So, we accept memes such as laws that punish thieves, thus acting to deter theft.

2.5   Memetics

Memetics is defined as the theoretical and empirical science that studies the replication, spread, and evolution of memes [12]. As part of a project with initial focus on establishing a scientific basis for memetics, Finkelstein compiled a broad overview of memetics with regard to human behavior and prospective military value [7]; it covers many written works by various authors on the topic. Related to replication, spread, and evolution, Finkelstein refers to concepts of information propagation, impact, and persistence with thresholds differentiated by orders of magnitude. This offered further quantification of the definition of a meme and its components toward establishing a scientific basis for memetics.

The first common model of memetics, as discussed above, is the biological analogy. In this model, memes are likened to viruses and genes. Memes can be considered as cultural equivalents of a virus that spreads via contact in a society. In the virus analogy, members who imitate each other have a passive role in the spread of memes in an environment. In the memes as genes analogy, evolutionary genetics is the main model and it carries over the ideas from biological models to memes.

The second common model of memetics is that of meme diffusion. In the book entitled "The Meme Machine," there is a view presented that people are merely devices operated by the numerous memes they host and are constantly spreading memes in a group [5]. Rosaria Conte argues an alternative view in which people are not vectors of cultural transmission, but rather actors behind this process [12]. Conte argues that dissemination of memes is based on intentional acting agents with decision-making powers [12].

2.6   Formal Definition of Memes

To facilitate a shared definition of a meme, a formal model of a meme was developed by Gunders and Brown [10], which is called the Gunders and Brown meme concept. In this concept, memes are made up of components and referred to as the Gunders and Brown Component Meme Concept, or CMC. The model is defined as follows:

$$CMC = < memetic\, engineer,\, hook,\, bait,$$
$$vector,\, host,\, memotype,\, sociotype > \qquad (2.1)$$

where the:

- memetic engineer is the creator of the meme
- hook refers to what attracts a person (or intelligent system) to the meme
- bait refers to what is the desired result promised by the meme to attract someone to consider it
- vector is the medium used to transport the meme to others
- host refers to the carrier that sent the meme

- memotype refers to the actual content of the meme
- sociotype is the social and cultural environment of the memotype.

These model attributes can be used not only to describe non-Internet memes, but Internet memes as well. Using the above "Kilroy Was Here" meme, the following formal CMC expression of the meme describes how it was initially spread during World War II:

Kilroy Was Here =

>*memetic engineer* =< *James Kilroy* >,
>
>*hook* =< *Who is Kilroy?* >,
>
>*bait* =< *Where Kilroy was* >,
>
>*vector* =< *WWII creates, inside ships, other places that were inspected* >,
>
>*host* =< *Ships, trucks and other freight transportation* >,
>
>*memotype* =< *Handwritten messages and picture* >,
>
>*sociotype* =< *Military personnel* > .

As the meme spread, the vector, hosts, memotype, and sociotype also changed. During its spread, the meme's vector changed from the inside of ships or freight to other vectors, such as print and electronic media and is now in granite on the World War II memorial in Washington, DC. The hosts changed from the ships to magazines and then eventually to the Internet. The memotype changed from just the handwritten note to also include the picture, and the sociotype changed from military personnel to the general population.

A formal model of a meme, as provided with the CMC formula, can help track how memes change over time, who changes it, which hooks and baits are more effective, how it spreads through and between sociotypes, and how the memotype and vector change as the meme spreads. The CMC formula is also useful for a community because it can be used to represent memes in a way that is meaningful and allows for its manipulation. A meme that simply exists in a space, like a picture of Kilroy, would instead have the capability to be engineered and modified if represented in a formal way. As discussed next, having a formal representation of memes also allows us to represent, manipulate, and share memes electronically.

2.7 Internet Memes

While Dawkins' original idea of memes describes memes as undergoing mutation by random changes and propagation by a form of evolutionary selection, memes can also be changed by intentional actors that make deliberate changes to promote a meme's fecundity. For example, an "Internet meme" is a subset of memes that can be mimicked or deliberately altered by human creativity [17]. Dawkins explains that Internet memes also have the additional characteristic of leaving a digital footprint in the medium in which they are shared so that they can be analyzed. The medium in which memes are generated and shared affects the way in which members of a community are able to use memes and shape the culture of a group. Digital memes have the unique ability to cause rapid changes in a culture as compared to traditional and analog memes because of their ability to be quickly mutated, consumed, and shared.

Popular Internet memes go through multiple mutations and are widely shared. The website "Know Your Meme" catalogs popular Internet memes by tracking their origin and displaying various user-submitted derivatives. For example, the "Business Cat" meme became popular in 2011 and started out as a picture of someone's cat with a tie on being posted to a bulletin board on the Internet [12]. Someone later added a background to the picture and put a saying on it that expressed something a boss would say followed by something involving what a cat does. From this version, people started putting a saying on the top of the picture expressing something one might hear from a boss, like "Get those reports to me," and then with a cat-related modification of the saying at the bottom of the picture like "right meow" [12]. A number of mutations of this were made over a short period of time, showing the speed that such memes can travel and the speed at which they can mutate. A few variations of this meme are illustrated in Fig. 2.2.

As discussed by Shifman [12], the Internet allows memes to spread quickly all over the world in short periods of time and reach a much larger audience. This contrasts with pre-Internet days when memes were shared from one generation to the next by word of mouth, writings in books, or other media. Pre-Internet memes spread and changed much more slowly due to the direct human to human nature of how they were shared. Examples of Internet memes range from cat videos, sayings such as "Leave Britney Alone," to pictures that are real or that are cleverly modified (i.e., "Photoshopped"). The Internet makes copying, modifying, and mimicking memes easy, and then people are able to share them quickly with others all over the world.

With the proliferation of smartphones, snippets of everyday life can now be easily recorded and shared with large audiences. When memes are sent over the Internet, they are stored on servers and individual computers all over the world. This allows them to be easily searched and saved indefinitely, giving digital memes an extreme

Fig. 2.2 Business cat memes

longevity and an audience that allows them to be rediscovered, modified, and updated based on the local culture or new time period, and spread once more, over and over again. Such memes can be constantly rediscovered and mutated making them useful or entertaining again.

Memes in a digital culture have higher rates of success because the three elements that a successful meme needs (longevity, fecundity, and copy fidelity) are positively impacted by the role of technology. First, regarding longevity, in a digital world information can be indefinitely stored and shared in a digital format unlike traditional memes. Second, regarding fecundity, the number of copies made within a time unit increased as digital communication makes it easier for spreading memes. Third, regarding copy fidelity, the accuracy in making new copies of memes has increased in digital formats because digital communication enables lossless information transfer [12].

The Internet also provides a medium for memes to compete with each other. With the ease of sharing memes on the Internet, too many memes become available for any person to view, so people become selective as to the ones they look at. The memes then have to compete to be the funniest, most informative, inspirational, or other factor of importance to people. The most popular and "viral" memes are those that are shared among Internet users and are remixed or modified to produce new memes. Because of the profusion of such memes, people with large followings on the Internet can have a major influence on the Internet and other cultures. How people filter such information and the information that they trust on the Internet, such as movie reviews, has been a focus of research [9].

Internet memes can quickly "go viral" and rapidly increase in popularity. Internet and social media communities, such as Reddit, Facebook, Twitter, or Instagram are platforms for memes to be curated and shared. On Reddit, the most popular memes are voted on by the community so that they receive more points and are pushed to the top of the popularity list—increasing their virality. Other platforms enable popular memes to spread by their members sharing a meme to their follower lists. Popular memes are then modified by users quickly over time so that the meme is applied to new contexts. Memes that are more versatile and that can be adapted to different contexts tend to remain popular for longer periods of time on these platforms [12].

Internet memes can also be described in a formal way as in the CMC formula developed by Gunders and Brown [10]. As an example, consider the "Grumpy Cat" meme that was immensely popular between 2012 and 2013 [13, 18]. This meme shows the Grumpy Cat with an unimpressed frown. In the image there is a phrase, like "Another terrible day in paradise" or "I had fun once—it was awful," which expresses an annoyed and grumpy mood. In this case, the memetic engineer would be the person who created the image or designed it (call the person "Fred"). The hook may be the facial expression on the cat's face and the clever associated text. The bait may be a description saying that the image will make you laugh. The vector is the Internet in this case, and the host might be Reddit or Imgur. The memotype is the actual image file that is uploaded to the host. The sociotype may be an image board that is associated with other humorous images which provides relevant contextual information to the viewer.

Using the above CMC formula, this meme can be written as:

Grumpy - Cat $=$

        memetic engineer $=<$ *"Fred"* $>$,

        hook $=<$ *"Grumpy Cat"* $>$,

        bait $=<$ *"It Will Make You Laugh"* $>$,

        vector $=<$ *Internet* $>$,

        host $=<$ *Reddit* $>$,

        memotype $=<$ *figure image* $>$,

        sociotype $=<$ *Humorous Imageboard* $>$.

If each meme had a formula that describes it, in a standard language much like Extensible Markup Language or XML, then we could look up what other memes Fred had engineered, or examine or search for other memes with the same bait, or join the sociotype that the meme came from. Other analyses that could be performed include tracking memes to see how they change over time (hook, bait, memotype), who is changing them (memetic engineer), and what changes cause memes to spread more rapidly (change of hosts and sociotypes). Such a structure for Internet memes would allow rich analysis of how information changes and flows through the Internet. Such information would be useful for marketers, advertisers, and product placement to allow them to keep track of what information works best for which sociotypes, which baits and hooks work the best for those sociotypes, and the memetic engineers that produce the most successful memes.

By analogy, adopting a formal description and standard language that gives memes the ability to perform similar analyses on their own, as well as modify themselves based on their own analyses, would endow memes with a life of their own. This is the topic of the next section.

2.8 Intelligent Memes

The Internet memes discussed above are static memes and need to be propagated and changed by people or machines. Due to the evolutionary pressures faced by memes, they can benefit by having the capability to promote themselves and provide for their own self-preservation. To this end, memes could be encoded with intelligence so that they leverage their flexibility in order to further their spread in a community based on that community's dynamic needs.

By adding intelligence to the memes themselves, the memes can become active entities in their own right. This means that the memes would have the ability to modify themselves to be more useful or to better spread themselves through their sociotype, or to even new sociotypes. A meme could modify itself through genetic or other evolutionary algorithms to test which modifications survive and thrive, or the meme could just observe how the environment is changing around it and actively evolve itself in parallel with its environment (see Fig. 2.3). Adding intelligence to memes would allow a meme to have a life of its own and not be dependent on outside entities to spread and modify it. By using their own intelligence, intelligent memes could actively compete with other memes and create more useful memes for their communities.

Intelligent memes would differ from ordinary memes in that they would have special intelligent agent-like capabilities that allow them to adapt to become more useful to a community, apply themselves to a problem, and interact with other memes. Intelligent memes would be able to change the values of their own parameters and knowledge bases to create new offspring memes. The offspring produced by a meme would be traceable back to the original meme for pedigree reference and to track its results. This would allow memes to produce better versions of themselves and for the new memes to become more effective at solving the current problems faced by a community.

For example, a robot's meme could adapt to the introduction of new robot hardware capabilities by generating meme offspring that have modified parameters and knowledge bases that are best suited for the new capabilities of the robot. Likewise, a new robot may have greater speed capabilities and would benefit from a new meme that executes certain steps at a faster rate.

Referring to the CMC meme model introduced above, *CMC = < memetic engineer, hook, bait, vector, host, memotype, sociotype>*, an intelligent meme could identify changes that could be made to variable parameters in the memotype to produce a different hook and bait based on its new capabilities. An intelligent meme itself becomes the memetic engineer because the intelligent meme is the creator of the new meme. An intelligent meme could analyze the sociotype of a meme to determine contextual changes to make to the memotype. A single meme may benefit from having

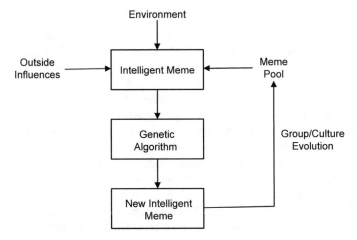

Fig. 2.3 Evolution of intelligent memes

slight alterations made to itself to adapt to different sociotypes. These changes could be performed by an intelligent meme itself, which would lead to faster evolution and learning in a community. An intelligent meme could make changes to the memotype in ways that would increase the fitness of a meme in a community so that the meme is more likely to be learned and used by members of the community.

An intelligent meme's ability to adapt is similar to how genes change over time in evolution. For memes, they adapt to maintain relevance. The updated meme becomes better fit for the changing vectors, memotypes, and sociotype to which the meme is applicable, and allows it to be shared more often. Like genes, memes that do not adapt, or memes that adapt in the wrong way, may not keep up with the changing world and environment in which they operate, and would eventually die out because they would no longer be useful and would not be shared. An intelligent meme could continue to evolve itself until it becomes useful again. Like evolutionary changes, the meme could change itself multiple ways ranging from small tweaks to the hook or bait to more substantial modifications to its memotype or sociotype.

Multiple intelligent memes could also cooperate or collaborate with each other to produce new memes. Intelligent memes could cooperate to produce new memes by examining their collective memes, combining them, weighting those that have a higher use, and then evolving the collective memes to produce a new generation of offspring. If the memes have multiple memotypes, sociotypes, and vectors, these could also be combined or tried with the newly generated meme to see which provides the meme with a more useful environment. The collaborative and active efforts of a group of intelligent memes will result in faster and more successful evolution when compared to individual efforts. This can be analogized to particle swarm optimization techniques that use the best performing members of a swarm as the basis for the next evolution of the swarm. As Kennedy and Eberhart describe in [11]:

> …particle swarm algorithm imitates human (or insects) social behavior. Individuals interact with one another while learning from their own experience, and gradually the population members move into better regions of the problem space…

Instead of swarms being represented as particles, the swarm, or collective, would instead be intelligent memes that are actively cooperating and collaborating to learn to produce better memes that are more useful to people (or the intelligent systems that they serve), and that operate optimally in their environments.

Intelligent memes also augment the capabilities of a robot by providing functional advice. For instance, when a robot is deciding whether to adopt a new meme, a robot would make this decision by performing its own self-evaluation. However, an intelligent meme could communicate with the robot to share information regarding its own self-evaluation so the robot makes a more informed decision. An intelligent meme could perform its own self-evaluation by analyzing historical data associated with its prior use and its successes. For example, the intelligent meme could predict its rate of success for a desired robotic task by studying its past use in similar situations.

An intelligent meme could also create new offspring memes based on the current or anticipated needs of the community. A new meme could be created on demand upon determination that there is an existing need for a meme that does not yet exist.

In addition, an intelligent meme could anticipate a future need of the community by analyzing current trends, future plans, and communicating with other robots to preemptively generate new memes that may be useful to a robot in the future. The offspring memes generated by intelligent memes could be simple or complex memes.

2.9 Memes as Ideas

Since regular memes can be thought of as representing ideas, intelligent memes, with their active nature, can be used as a form of artificial intelligence for computer systems. Memes provide knowledge to the computer system through its content. And with the meme's self-adaptation through its active changing of its content to increase its usefulness, the computer system will exhibit intelligence through this active and autonomous updating of the system's knowledge.

Since memes can be thought of as ideas, they can be represented mathematically according to concepts described in Ulf Grenander's Calculus of Ideas [8]. In the Calculus of Ideas, ideas are represented in their most primitive form as "generators". These generators are connected to other generators through inbound and outbound connections. These connected primitive ideas form complex ideas. The connections between the ideas form a graph structure which can be analyzed. The ideas then make up thoughts and the thoughts form patterns. The thoughts can be complete (all inbound and outbound connections are connected) or incomplete (if there are any inbound or outbound connections to an idea that are not connected). This construct enables many possible operations and manipulations, such as comparing ideas to determine similarities and differences, forming abstractions of ideas to form higher-level ideas, and collecting ideas into groups to form concepts.

Based on the above short description of a concept from the Calculus of Ideas, there can be seen some similarities to memetics. Also, like ideas, memes can be linked together to form more complex memes. For example, the approach to mapping out the manner in which a meme is formed from other memes is similar to Grenander's illustration of how ideas can be mapped out. In this vein, Fig. 2.4 illustrates the combination of two memes (cat and business person) to form the business cat Internet

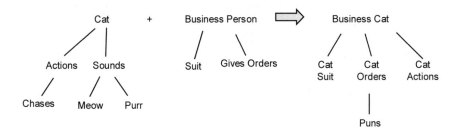

Fig. 2.4 Cat and business person meme combined to produce business cat

meme described earlier. The business cat meme, as described earlier, follows a format where an image of a cat wearing a tie is accompanied by text that blends business expressions with cat-related themes and puns. For instance, the example provided in the Internet memes section included a picture of the business cat with the text "get those reports to me right meow."

Intelligent memes can be represented through the calculus of ideas, or similar calculus. Basic memes, m, are combined using a meme generator, σ, to form complex memes with $M = \sigma(m_1, m_2, \ldots, m_n)$ where each of the $m_i \in M$ is a meme with connections to other memes in the list and M is a set of all possible memes from the collection of m_i. In the simplest case, there is only a simple meme in the list by itself. Connections between memes could also be defined, where each meme has defined inbound connections and outbound connections [8].

A number of operations on the set of memes, M, could be used to provide information on the set, or comparisons between sets of memes. A function for similarity, S, could check how similar two sets of memes are to one another.

In the Calculus of Ideas [8], ideas were mapped out based on their connections and relationships to other ideas. Memes could be mapped out in the same way to provide a visual (as well as computational) mechanism to view how a meme is constructed and how it compares to other memes. These maps can be represented in a mathematical way which allows them to be analyzed. They also offer a window into the "brain" that allows interpretation and explanation by humans, which can facilitate understanding and management of behavioral expectations, not to mention, trust, through transparency. A collection of ideas can be analyzed to determine the personality of someone (or something) who contains those ideas; the ideas can be analyzed as well to find flaws in thinking or patterns of thought and to determine if various pathologies exist.

2.10   Conclusion

Beginning with the origins of a meme, we have provided an overview of memes and explained how developments in technology have impacted memetics. Additionally, a formal model for memes was introduced that is further developed in later chapters. Looking beyond static memes, we introduced the novel concept of "intelligent memes" which gives memes the capability to self-modify. Lastly, we explored the possibility that memes could be used to form ideas and constructions that can be visualized to allow for interpretation and explanation of decision making and intelligence. These elements enrich the foundation and potential for robot memes that will contribute to next-level robotic intelligence.

References

1. Adleman L (October 15, 2014) Resurrecting smallpox? Easier than you think. The New York Times. https://www.nytimes.com/2014/10/16/opinion/resurrecting-smallpox-easier-than-you-think.html
2. Adleman L (July, 2018) Genes, memes and cenes: a general theory of evolution. Draft. https://cpb-us-e1.wpmucdn.com/sites.usc.edu/dist/4/121/files/2018/03/20180722-prenes-excerpt-2052hym.pdf
3. Agapakis C (June 14, 2014) Genes cannot be made from scratch. Scientific American. https://blogs.scientificamerican.com/oscillator/genes-cannot-be-made-from-scratch/
4. American Heritage Dictionary (2006) American Heritage Dictionary of the english language, 4th edn
5. Blackmore S (2000) The meme machine. Oxford University Press, USA
6. Dawkins R (1989) The selfish gene. Oxford University Press
7. Finkelstein R (May 2008) A memetics compendium. compiled by Robert Finkelstein. https://robotictechnologyinc.com/images/upload/file/Memetics%20Compendium%205%20February%2009.pdf
8. Grenander U (2012) A calculus of ideas: a mathematical study of human thought. World Scientific
9. Golbeck J (2008) Trust on the world wide web: a survey. Found Trends Web Sci 1(2):131–197. https://doi.org/10.1561/1800000006
10. Gunders J, Brown D (2010) The complete idiot's guide to memes. ALPHA
11. Kennedy J, Eberhart R (1995) Particle swarm optimization. In: Proceedings of the fourth IEEE international conference on neural networks, Perth, Australia. IEEE Service Center, pp 1942–1948
12. Business Cat Meme (2016) Know your meme, literally media. https://knowyourmeme.com/memes/business-cat
13. Grumpy Cat Meme (2016) Know your meme, literally media. https://knowyourmeme.com/memes/grumpy-cat
14. Shifman L (2014) Memes in digital culture. MIT Press, Essential Knowledge Series
15. Wade N (May 20, 2010) Researchers say they created a 'Synthetic Cell'. The New York Times. https://www.nytimes.com/2010/05/21/science/21cell.html?_r=0
16. Kilroy was here November (20, 2018) Wikipedia. https://en.wikipedia.org/wiki/Kilroy_was_here
17. Internet meme (November 20, 2018) Wikipedia. https://en.wikipedia.org/wiki/Internet_meme
18. Grumpy Cat (November 20, 2018) Wikipedia. https://en.wikipedia.org/wiki/Grumpy_Cat

Chapter 3
Robot Memes

3.1 Introduction

Memes can be thought of as the genes for robots. They can be used as a mechanism for sharing information between people, between machines, between people and machines, for describing a culture, representing shared knowledge, ideas, and intelligence. Memes can also be used as a basis for sharing ideas and intelligence between robots. Memes can be used as an active knowledge base that can endow robots with enhanced and higher-level intelligence, adaptation, self-awareness, and the basis for culture in a community of robots. Like regular or intelligent memes, robot memes can be modified to allow a robot or group of robots to explore new ideas and can be used for discovering new knowledge, ideas, and ways for robots to perform their duties.

In this chapter, the ideas presented in earlier chapters are adopted and extended to apply to intelligent robots. First, previous research related to robot memes is briefly discussed for appreciation of prior thinking and to establish any influence on related ideas presented herein. This is followed by a description of some contrasts between robot memes and regular memes discussed earlier with humans in mind. A formalization of robot memes is then introduced to inform considerations of their practical application. Armed with this background, the manner in which memes and memetics can help to form a culture among robots is explained. Lastly, an approach to using robot memes not only as a basis for robot knowledge but as a basis for robot intelligence is described.

3.2 Defining a Robot Meme

Using memes as a basis of robot knowledge has been discussed by several researchers [2, 3, 5, 8, 9]. Feng described the use of memes in robots as follows:

© The Author(s), under exclusive licence to Springer Nature Switzerland AG 2020   31
W. Truszkowski et al., *Robot Memetics*,
SpringerBriefs in Electrical and Computer Engineering,
https://doi.org/10.1007/978-3-030-37952-0_3

The instructions for carrying out the behavior to act on a given problem are modeled as knowledge memes. The knowledge memes serve as the building blocks of past problems [sic] solving experiences that may be efficiently passed on or replicated to support the search on future unseen problems, by means of cultural evolution. This capacity to draw on the knowledge from previous instances of problem-solving sessions in the spirit of memetic computation…thus allows future search to be more efficient on related problems. [3, p. 162]

According to Winfield, a robot meme is:

a contiguous sequence or package of behaviors copied from one robot to another, by imitation. [9, p. 261]

In the following, we expand on these definitions and provide a structured meme that can be used by robots to communicate ideas and knowledge between themselves and humans. This structured meme can form the bases of robot intelligence, adaptation, and the ability to form complex societies and culture not only with other robots, but with people as well.

Allowing robots to choose what learned behaviors to enact gives rise to behavioral evolution which allows adaptation to new environments, to other robots, and to humans. Winfield describes the concept of a robotic culture as measurable and sustainable differences in the memes across different groups of robots, where those memes can be traced back to common ancestral memes. We would add to this that the memes need to be intelligent memes as described in the previous chapter.

3.3 How Robot Memes Differ from Human Memes

We define *robot memes* as learnable real-world patterns or knowledge encoded in computational representations for the purpose of effective problem solving. Robots can possess special agent generation capabilities so that when a robot sees a meme, it will be able to translate its sensory observations into a form it can understand and act upon.

A meme can also be embedded with historical information that can be analyzed by agents such as past intended function, environmental conditions of prior use, mutation history, and sharing statistics. An agent can analyze this information in deciding whether to learn a meme, determine how to mutate a learned meme, or discard it.

In agents, examples of memes include algorithms, observations, knowledge, instructions, actions, or ideas (Fig. 3.1). For robots, memes can also represent actions involving mobility or manipulation, navigation from one place to another, or the associated behavior or skills that they enable (e.g., associated with construction, maintenance, assembly, repair, etc.). Robot actions represented by memes may also include sensing, perception, planning, and reasoning actions involving actuation. Movement could also apply to mobile software agents that move from one computer to another.

As discussed in the previous chapter, memes can be simple in the sense that they can encompass a single idea, or represent a complex idea where a meme is comprised

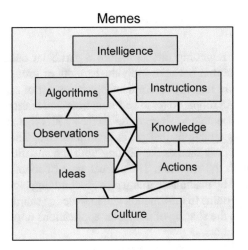

Fig. 3.1 Context of memes

of multiple simple memes that are related to each other. For robots, simple ideas can form individual actions or observations of the world while complex ideas are a collection of related actions or observations. For example, a sequence of tasks could be represented as individual ideas that form a more complex task, such as tasks of walking, bending, and picking up an item could be combined to perform the complex task of clearing an area of debris. Like complex memes, complex ideas can also be formed through a natural selection process, such as a genetic algorithm, that combines simple ideas to produce new complex ideas that did not previously exist until the need arose, such as cleaning up an area after a storm versus normal cleaning. Complex ideas can be gradually changed and updated over time as tasks and observations become more complex. Complex ideas can also make it easier to modify and evolve new problem solutions since they have more information embedded in them.

Memes can also be used to act as an exchange of cultural information or ideas for robots and intelligent agents. Robots may be working on individual tasks or working as part of a group to accomplish a job and may need to communicate with other robots or people. The robots would communicate with each other through some defined robotic language to exchange information. Robots are also able to observe the environment and each other in real time within the environment they are working. The robots can plan their activities and make real-time adjustments based on unanticipated and possibly harmful circumstances. This information can be communicated between the robots through memes.

3.4 Robotic Meme Communication

As discussed earlier regarding memes within a particular culture, memes can be replicated and transmitted between individual robots or groups of robots. This is also a desirable feature for robot memes enabling transfer of capabilities or knowledge among robots. As robots adapt to new situations and learn or otherwise acquire memes, they should be able to share those memes with other robots that later find themselves in similar situations that are at first unfamiliar. Those robots should be able to adopt appropriate memes for dealing with the situation from other robots that are familiar with the situation. The current state of robotics development and capabilities exhibited by machine learning provides building blocks that address similar capabilities as pertains to information and knowledge sharing among networked robots. This includes the sharing of computer applications or programs and learned capabilities among robots.

Meme transmittal or transfer between robots is akin to skill transfer and transfer learning concepts of intelligent robotics and machine learning and, more generally, of intelligent systems. Robot memes may be transferred horizontally, unless copied via "whole-brain" transfer (as a total software image, i.e., akin to vertical transfer from a robot's "parent" or, in the context of the CMC model, the memetic engineer or host). In the broader study of memetics, meme fitness has been suggested to mean ease of transfer [4]. If applied to a robot meme, this may imply that such memes should be easy to receive, and that the recipient robot's sensing and acting capacities should be compatible to those of the donor robot; otherwise, a meme may not be usable. For example, consider a meme to rub one's stomach triggered by sight or smell of delicious food. Such a meme would not be (easily) usable by a recipient without the sense of sight and/or smell. Alternatively, meme fitness suggesting ease of meme transfer may imply that such robot memes "must be generalized into a common representation" [7], which would facilitate transfer in a pragmatic and open architecture sense.

Related to the concept of meme transmittal/transfer is the sharing of robot or general computing resources with other robots via access to a "cloud" infrastructure (i.e., cloud robotics) [6]. In some cases, robots are connected via a vast network to a repository of shared information that facilitates or accelerates learning about their own behavior or environments and provides for more computing power. Robots so connected can also contribute their own data/information and knowledge to the same shared repository. Repository content may come from a variety of sources such as robot experiences while operating in the real world, computer simulations, the Internet, and information/knowledge provided by humans. Evolution of this knowledge sharing among robots will facilitate the replication and transmission of robot memes in a given culture.

Similar to cloud robotics, robotic memes could be stored in a meme cloud and used for future robotic actions (Fig. 3.2). Memes should be thought of as active entities that are constantly undergoing change in an evolutionary fashion to adjust a robot's behavior over time. The memes can be modified by the robot's current base of

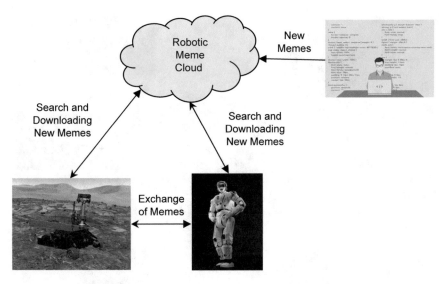

Fig. 3.2 Meme cloud shared between robots. Image credits Courtesy NASA/JPL-Caltech, NASA, pixabay.com

observations and models, and merged with existing memes to evolve a robot's memes using outside inputs. These memes can then be shared with other robots that will affect their behavior as well as the behavior and the culture of the robot community.

Knowledge can be readily transferred within a robot community through a pool of shared memes (Fig. 3.3). The meme pool represents the collective problem solving experiences of the community. This rich history can be shared with all of the members of the community enabling the robots to make evolved decisions as the collective experience grows. The meme pool enables a robot to seek a specific meme from the pool or directly transfer a meme to another robot.

Additionally, memetic knowledge can be transferred directly through observation and imitation. Knowledge gained by observation is a separate way of learning than learning new memes by tapping into the meme pool. Learning through observation requires robots to actually observe a behavior in real space in order to later imitate the learned behavior.

3.5 Robot Meme Initialization

Though a robot's memes can evolve over time, a robot would need to start with a base, or initial set of memes that allow the robot to operate in at least a nominal way in its target environment. Providing a robot with a richer set of memes would allow a robot to more easily create new memes from modification of a larger set. This set should at least contain the memes that would allow a robot to perform its

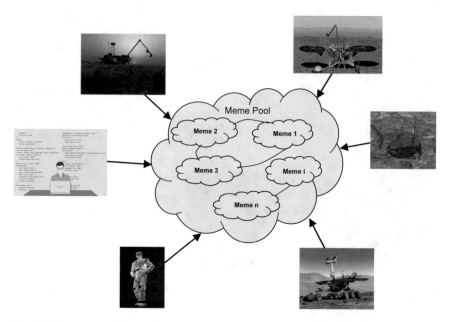

Fig. 3.3 A meme pool where robot memes can be shared. Image credits Courtesy NASA/JPL-Caltech, NASA, pixabay.com

intended task; this would allow it to be productive right away. If a robot's task is not well-defined, or its environment is not well-known, a supervisor could start a robot with a base set of memes that are the best guess of the types of memes it will need given the tasks it is to do and its envisioned environment. This set should be at least a base set of memes that will allow it to operate at a basic level of functionality and build on over time based on its interaction with other robots and its environment, and via learning and reasoning about its environment. If the base set of memes is not sufficient, it can modify this set of memes through evolutionary or other processes, through observation of other robots, exchanging memes with other robots, retrieving memes from a central repository, or receiving new memes from the humans or agents supervising the robot.

3.6 Memes and Robot Culture

The environment in which the robots perform their work gives rise to the culture of those robots. The culture here, in the robotic context, arises and evolves from the individual and collective behaviors of the autonomous intelligent agents who "live" and "work" in a community. Memes give the community of autonomous entities more power and knowledge to accomplish their collective tasks through a shared culture. There are still many unknowns regarding the representation and transmission of memes in an agent community, which is a motivation for this exposition and a basis for future research.

Robot culture can be used to describe the cultural communications between robots as well as between robots and humans. As robots become increasingly intelligent they may, as a robot community, generate societal norms that describe the general customs or way of life. In order for robots to develop a culture, robots would be required to possess the capability to be self-aware and self-conscious. These requirements (toward which progress is being made [1]) may also be incidental to being capable of performing memetic learning.

Robot culture may also be shaped by human and robot interactions. Values and norms that exist in human culture may find a way into robot culture through interactions between humans and robots over time that involve meme exchanges. This relationship between humans and robots may have bidirectional effects on both human and robot culture. For example, humanoid robots may adapt to changes in the environment differently after observing humans performing similar tasks. Alternatively, humans may be inspired to create new designs and algorithms after observing solutions generated by robots.

Memetic learning enables robots to learn social values by performing intelligent imitation. A robot community will be unable to propagate useful social values if the members of the community indiscriminately learn every observed behavior. Intelligent imitation is performed when robots can decide whether an observed behavior is better than the current existing behavior [8, 9]. Robots can analyze factors such as efficiency, risk, time, and success to determine whether a new observed behavior is worthwhile. If an observed behavior would improve its own behavior, then a robot can incorporate another robot's behavior through imitation. The robot may change or improve on another robot's behavior over time.

Humans can also impact the robot community by preloading robots with memes that guide cultural values. Future modifications and learned behaviors can be compared against a set of idealized values that are preloaded in the robot community. An example of such a meme is to avoid damaging other robots in order to promote selfish gain or to avoid adopting policies that destroy the infrastructure to accomplish a task. These guidances can be initially loaded and also injected over time as the community evolves and demonstrates new functions.

In addition, a robot community can be guided toward certain social values and norms by injecting memes into their community. For example, humans can monitor the state of the robot culture and decide that a certain social policy needs to be adjusted for the well-being of the community. Then, memes that act to coach the robots toward a different direction can be shared with the community. The robots will understand that those memes should be valued because they are encoded with information reflecting their importance. Humans can also guide the community by removing memes from the community or modifying the contents of existing memes. Such ideas foster a means for ethical design of intelligent robots.

3.7   Using Intelligent and Complex Memes for Robot Intelligence

Intelligent memes, as discussed in Chap. 2, have their own intelligence built into them and can promote themselves and provide for their own self-preservation through self-modifications based on the changing environment and the needs of the robot community as well as through spawning of new memes that may be useful. The memes can also make sure that they are being used in an ethical manner based on their built-in ethics models. Intelligent memes can also be chained together to form complex memes, used to represent knowledge the robot has and ideas on how a robot might solve new problems it encounters. These ideas can be explored through simulations, mathematical analysis as well as trial-and-error techniques to refine them.

An example of a complex meme is illustrated as a graph structure in Fig. 3.4. In this example, a number of memes have been chained together both linearly (like sequential steps to take to complete a task) as well as hierarchically as a decomposition of higher-level concepts or tasks into lower-level ones. In this example, a robot is trying to move a communications tower that has fallen across a road during a storm. The need to remove the tower is represented as an idea. The idea contains the steps needed to remove the tower from the road and put up a temporary communications tower. The horizontal tasks represent a linear sequence. Each of the tasks are then decomposed into its constituent subtasks, and sub-subtasks where needed. Each of the tasks shown in Fig. 3.4 are still at a high level, but would be broken down into detailed steps to execute the task. The idea could be simulated to predict the likelihood of success and to determine if the idea is sound or not. The memes that

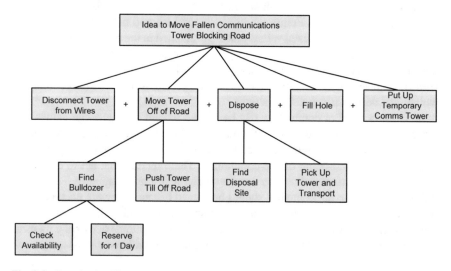

Fig. 3.4 Complex intelligent meme for moving a communications tower that is blocking a road

make up the tasks could also be self-organizing. If each higher-level task has a set of pre-conditions and post-conditions, then the memes could self-organize by ordering themselves so that given the preconditions, the actions inherent in the memes (once executed together as a complex meme) fulfill the post-condition. Gaps between post-conditions and pre-conditions could represent holes in the memetic sequence.

If memes are not available to implement all or part of the idea, the robot can retrieve memes from the meme pool containing memes in the broader robot community, download appropriate memes from the meme cloud, or run the memes through genetic algorithms until a meme that produces the desired actions is evolved. Additionally, a supervisor, such as a human or supervisor agent, can also inject memes into the meme pool until the desired memes are found. In this way, the robots can continuously evolve.

The intelligent memes enabling robot intelligence can be thought of as autonomous, autonomic entities. They are constantly looking to try to survive and will go through evolutionary transformations to survive and stay useful.

3.8 Bad Memes

Memes may not always be beneficial to the community. They can be malformed, corrupt, malicious, or, through evolution, genetic algorithms or mutations can end up with information that causes a robot to become a bad actor or just represent incorrect information. A robot that is a bad actor could be executing a meme that causes it to operate incorrectly causing problems instead of solving them. This could cause the bad actor robot to interfere unintentionally with other robots, or even cause it to be confrontational with other robots in its group, preventing them from being able to complete their work and affecting the culture of the robots.

Over time, bad memes could also evolve into memes that are even worse. These bad memes could be malformed based on a misinterpretation of another robot's actions, evolved incorrectly through a genetic algorithm or incorrectly injected by a human. Malicious memes could also be purposely injected into a robot, or robot community by a human or a rogue robot, much like a computer virus, and represent a cybersecurity risk. Malicious memes could disrupt robots, or try to change a robot culture to align in negative ways preferred by the malicious actor.

External manipulation through injection or coaching can amend incorrect or malicious behavior. Injecting corrective memes could allow the robot to learn a corrective meme that replaces the bad meme. Circular memes could also develop over time. These could be a series of memes to perform a task where one task ends up forming a loop and a task never gets completed. Repetitive tasks could also form a meme loop; so analysis on loops would need to be done to make sure they are properly designed to achieve the desired outcome.

Intelligent memes can also possess the ability to identify when corrective action is needed by analyzing the results of their own actions. One way that intelligent memes can identify when their behavior requires corrective action is by comparing its current implementation with its past history. This type of intelligent version control or regression test could be used to identify cases when a bad or malicious meme has been created through corruption or misinterpretation. An intelligent meme could also use internal ethics that would identify a bad meme and prevent it from being used, modified, evolved, or shared with other robots. Similarly, a robot can be coached by another robot or human to correct its behavior by demonstration or by instructing it to no longer use its bad meme.

When a robot observes another robot performing what it considers to be a bad meme, there could also be a reporting feature which enables further attention and examination to be drawn toward the reported behavior. This way, the community can shape the collection of memes by passing judgment on observable behavior. For example, a meme that receives multiple reports within a period of time can trigger some automated actions, such as disabling the meme, which may limit its perceived negative impact, or even having it self-destruct. Since memes can become viral and spread quickly within a community, it is important to have community-based moderating which lessens the effect of bad memes. A meme vaccine could also be produced that counteracts bad memes and makes them ineffective and inhibits their being passed along. A meme vaccine could be generated after a bad meme is observed, or a vaccine could be generated that could inoculate robots against a class of bad or malicious memes, similar to the way in which cybersecurity anti-virus software is used now.

3.9 Conclusion

This chapter discussed the use of memes for robot intelligence and for forming a shared culture with other robots, agents, and humans. The memes form the basis of the information, ideas, and knowledge exchange between entities to increase the group intelligence and the ability for robots to perform new tasks. Memes could be shared using a meme cloud, similar to cloud robotics, where robots can search a cloud for needed skills, behaviors, or other useful information associated with a given robot culture. A robot could be initialized with a set of memes that over time would be self-modified based on the situation and environment that the robot was placed in. As the robot spends time in the environment, it would expand its collection of memes based on the knowledge of other robots in the community, the tasks it is performing, and generation of new memes to accomplish new tasks. This memetic learning would also allow it to learn social values by intelligently imitating other robots in the community. It is also possible that memes that are generated or learned may not be beneficial to the community due to being malformed, or just conveying bad information, which could cause a robot to become a bad actor. These bad memes could also spread to other robots if they are not noticed by others and reported, so constant observation of the robots by each other is needed to catch a bad meme as soon as possible. External coaching or injecting of new memes are possible mechanisms that could be used to correct or override bad memes and make the robot a useful member of the community again. Armed with robot memes and a capacity for memetic learning, one can begin to consider how the notion of memetic algorithms might contribute to problem solving abilities for intelligent robots.

References

1. Bringsjord S, Licato J, Govindarajulu NS, Ghosh R, Sen A (2015) Real robots that pass human tests of self-consciousness. In 2015 24th IEEE International Symposium on Robot and Human Interactive Communication (RO-MAN), Kobe, Japan
2. Feng L, Ong Y-S, Tan A-H, Chen X-S (2011) Towards human-like social multi-agents with memetic automaton. In: 2011 IEEE congress of evolutionary computation (CEC), pp 1092–1099
3. Feng L, Ong Y, Tan A, Tsang I (2015) Memes as building blocks: a case study on evolutionary optimization + transfer learning for routing problems. Memetic Comp. 7:159–180. https://doi.org/10.1007/s12293-015-0166-x
4. Higgs PG (2008) Is it good to share? The parallel between transfer of information via memes and horizontal gene transfer. In: Symposium on memory, social networks, and language: probing the meme hypothesis II, University of Toronto, Toronto, Canada. May 15–17, 2008
5. Hougen DF, Carmer J, Woehrer M (2003) Memetic learning: a novel learning method for multi-robot systems. Robotic Intelligence and Machine Learning Laboratory, School of Computer Science, University of Oklahoma, Norman, OK
6. Kehoe B, Patil S, Abbeel P, Goldberg K (2005) A survey of research on cloud robotics and automation. IEEE Trans Autom Sci Eng 12(2)
7. Meuth R, Meng-Hiot L, Ong Y, Wunsch D II (2009) A proposition on memes and meta-memes in computing for higher-order learning. Memetic Comp 1:85–100
8. Vanderelst D, Winfield AFT (2016) Rational imitation for robots. In: Tuci E, Giagkos A, Wilson M, Hallam J (eds) From animals to animats 14. SAB 2016. Lecture notes in computer science, vol 9825. Springer, Cham. https://link.springer.com/chapter/10.1007/978-3-319-43488-9_6
9. Winfield AFT, Erabs MD (2011) On embodied memetic evolution and the emergence of behavioral traditions in robots. Memetic Comp 3:261–270. https://doi.org/10.1007/s12293-011-0063-x

Chapter 4
Memetic Algorithms

4.1 Introduction

Memetic algorithms enable robots to engage in memetic learning—that is, the transmission of cultural ideas or symbols from one robot mind to another, enabling robots to learn social values by performing intelligent imitation. We adopt the definition of Ishibuchi to describe memetic algorithms as:

> A synergy of evolutionary or any population-based approach with separate individual learning or local improvement procedures for effective problem solving, based on Darwinian principles of natural evolution and Dawkins [sic] notion of a meme. [1, p. 75]

In this chapter, we discuss the minimum set of functional requirements a robot should have in order to process memetic algorithms and perform memetic learning. While the listed requirements below are not exhaustive, a few of the important functional requirements and considerations include the following: observation, self-awareness, evaluation, and manipulation. Each of these is elaborated below.

In addition to robot requirements for memetic learning, we discuss the formalization of robot memes and how they can be used in problem solving. The formalization is based on that of an intelligent meme as presented in Chap. 2 with information added to support robot memetic algorithms. An example of robots working in an uncertain environment is given to help ground the formalisms developed.

4.2 Observation

The first requirement to perform memetic learning is the ability to observe memes. Although robots can have the capability to communicate with other robots and can access a common repository of memetic information, it is essential that robots should also be capable of observing another actor performing a task to learn new memes.

© The Author(s), under exclusive licence to Springer Nature Switzerland AG 2020
W. Truszkowski et al., *Robot Memetics*,
SpringerBriefs in Electrical and Computer Engineering,
https://doi.org/10.1007/978-3-030-37952-0_4

A robot should be able to observe another robot's performance of a task and derive a set of memes from observing the task and mimic either all, or most elements, of the observed behavior. For example, a robot can visually observe another robot climbing the face of a cliff to learn new memes about climbing methods, terrain, or component durability.

Robots could observe both other robots and humans to learn new memes by observation. Depending on the complexity of the meme and the context of the observation, a robot may observe and learn only parts of a complex meme. Robots and humans can also engage in a teaching function where they can visually teach other robots new memes. Likewise, a robot could request that another robot or human teach them a new meme by performing a task. This functionality also leads to more collaboration between members in a robot society and advances robotic culture.

Robots that can observe memes should have the ability to utilize computer vision to identify what memes are being used and (or may apply) to try and understand the intention of the actors. Robots should have the capability to use the data collected and derived from their sensory inputs and knowledge to gain a high-level understanding of their observations. A robot's observations about another's intention behind executing a behavior will be affected by the robot observer's frame of reference. Without full context provided by communicating directly with the actor, the observer may have to make assumptions and inferences to understand the intention of the actor. For example, a robot observing another robot climbing a cliff face may assume that it is doing so because scaling the cliff is the most efficient way to reach a needed resource. However, the observer may not have been aware that the actual reason the robot is climbing the cliff is because the robot had earlier stumbled onto the cliff face from above and is only trying to recover and avoid falling to the ground below. Although the observer is not able to determine an accurate intention to give factual context to the observed memes, the variation in new memes generated by differences in frame of reference may promote a diverse meme pool and creativity to solve new problems.

4.3 Self-awareness

Robots must be conscious of their state and situational task or mission context in order to perform memetic learning. To effectively solve problems, robots must be able to recognize that they have encountered a problem and set individual goals to solve it. One way that a robot can be aware of a present or anticipated problem is by developing and evaluating a self-model in order to track its state and understand the effects of its actions. When a robot is able to track and recognize its current state and situation, it will be able to comprehend the effects of actions it is performing and track progression toward its goals. By monitoring its own state over time, a robot can realize if there is a problem because it can compare its current state to a goal, simulation, or historical information of a meme.

Also, if a robot has a more sophisticated level of awareness, then it can also understand and explain why it is performing a certain task. Robots can also reach a level of social awareness so that they not only understand why they are doing a task, but can make the same analysis for other observed members of the community. By understanding on its own what a robot is doing and why it is doing a present task, a robot can begin to develop individual as well as community goals. This level of human-like consciousness enables the creation of a complex culture and learning.

4.4 Evaluation

When a robot learns a new meme, either through evolving a current meme or through an outside source, the robot will need to determine if the new meme will be able to accomplish the given task. The robot may also need to evaluate several different memes to determine which one is best to complete a task. In order to compare available memes, a robot should be able to compare its memes to memes in the meme pool by analyzing parameters from memetic expressions. Depending on the robot's scenario, different parameters of a meme can be prioritized and evaluated. For example, it may be a requirement for a robot to seek a meme that promotes speed over power consumption to accomplish its task. Based on this requirement, memes with parameters that promote speed will be scored higher during evaluation.

A robot can also judge the effectiveness of memes through exploratory trial and error. However, there may be times when the robot may only have a limited number of tries to accomplish a task, such as if there is a limited resource involved or the nature of the operation can only be done once. In these cases, the robot may need to *simulate* the actions of a new meme or current memes before actually using them. Meme simulation could reduce the number of physical trials a robot needs to execute before finding the right meme. Executing a meme in simulation could be much faster than experimenting in the real world, which may allow the robot to learn faster. This could be important in time and life critical situations where the robot is not able to try a number of memes to find the best one. The simulations may not be perfect, in that the meme may work in simulation and not in the real world, but using simulation to predict what the meme(s) would do first would increase the chances that the chosen meme is the correct one.

Coaching could also be used as a form of memetic engineering for robots and agents. Coaching would involve offering changes to current memes that are being used that could make them more efficient or updating them for new tools or tasks. The coaching could be done by humans, other senior robots, or specialist software agents. Robot coaching could be used not only to inject new memes or modify current memes; it could also be used to help evolve a robot community toward a particular direction by shaping the pool of memes in use by the robot community.

Injecting memes into or coaching a group of robots could also be a way of getting a group of robots out of a "groupthink" mode or a way of assisting when they are unable to solve a problem. If the meme pool becomes stagnant over time, the robots may continue trying the same unsuccessful actions or may cease the process of evolving and getting better. New memes could be injected periodically by outside entities, or when the memes are no longer evolving. These new memes could be developed by a supervisor through genetic algorithms. Measuring how much a meme pool is changing could indicate whether the robots are in a groupthink mode or not. If little or no change is occurring, then the pool needs to be modified.

4.5 Manipulation

Existing memes in the meme pool will not provide robots with the knowledge to solve unexpected problems. Robots must be able to manipulate existing memes to generate modified memes in order to solve new problems. Memes can be evolved by the original host, or by those who seek to imitate the behavior. A robot may decide to modify an existing meme when it determines that the meme is not able to accomplish its intended behavior because of the robot's new environment and/or context.

Memes can be modified by genetic algorithms which tweak parameters in the memetic expression to produce new offspring. Different memes can even be combined together to find solutions to more complicated problems. When a combination of two different memes occurs, traits from multiple memes are combined to generate a new meme which has desired characteristics from its parents. The combination of different traits of memes can form a complex meme which has a variable influence from its constituent memes. In this combination, a combined complex meme has characteristics from different memes in order to form a new meme and to perform a new function.

If the existing memes do not provide enough knowledge for a robot, then a robot can conduct mutations to a meme in order to find a new solution to its problem. Performing mutations to a meme can be based on randomness or by using simulations to optimize the trial-and-error process. As in evolutionary robotics, results from simulations can be used in conjunction with genetic algorithms or genetic programming to simulate the new memes and determine if they are useful or not.

4.6 Formalization of Robot Memes

The formal definition of a meme given earlier (in Eq. (2.1) of Chap. 2, Sect. 2.6) can also be used as a formal definition for transferring memes between agents and robots, either communicating with each other or as a way for people to transfer information to autonomous memetic systems. Since the agents and robots that will be exchanging the memes are autonomous and autonomic systems, the meme should contain utility types of information instead of sensational, entertainment, or sales types of information. The agents and robots are more likely to accept a meme if it can potentially solve a current or future problem.

A formal definition of a Component Meme Concept (CMC) that could be used in robotic communication is the following

$$CMC_R = < meme\,name,\, creating\,robot,\, subject/problem,\, results,$$
$$network,\, sending\,robot,\, meme\,content,\, robot\,community >$$

$$(4.1)$$

where

- The *meme name* is a name for the meme that can be referenced
- the *creating robot* is the same as the memetic engineer (the entity that created the meme)
- the *subject/problem* replaces the "hook" that attracts other robots to the meme
- the *results* of the meme serve as the bait and describe the results of using or applying the meme
- the *network* is the medium through which the meme is being transmitted
- the *sending robot* is the host robot that sent the meme, which may be different from the creator of the meme
- the *meme content* is the memotype and can be some or all of the components shown in Fig. 3.1 of Chap. 3.
- the *robot community* is the sociotype, which is the community wherein the creating robot was situated when the meme was created.

Additional theories can be applied to supplement formal robot meme definitions by offering a cognitive robotics foundation based on formal mathematics. This is exemplified in the behavior engine of an architecture, called CARACaS (Control Architecture for Robotic Agent Command and Sensing), developed by NASA [2] to enable intelligent robotic systems for future space and planetary missions. That behavior engine has as its mathematical basis a formal process algebra, in particular, a cost-calculus framework [3]. Approaches based on concept algebras such as denotational mathematics [4] could also be considered.

The following section expresses how a meme might be created and how it might flow through a community of agents/robots.

4.7   Robot Memes Used in Problem Solving

To appreciate how a community of robots would work together, learn, and cooperate, and how the robot community culture would change when memes are shared between robots, consider the example of a humanitarian assistance and disaster relief (HADR) scenario after a hurricane. In a robot-assisted disaster relief effort, there may be robots that move debris, clear roads, search for survivors, transport humans, and set up shelters and other temporary infrastructure, such as communications antennas. The robots may also take many forms, such as robotic bulldozers, cranes, humanoid robots, robotic vehicles (e.g., busses, cars, drones), and other specialized robotic forms. There could also be robots in a command and control (C2) "center" to organize the effort, which may be distributed over the disaster area and even around the world when there are sufficient communications.

The "robots" in the C2 center could be autonomous software agents rather than physical robots. For C2, planning, logistics and other functions not requiring motion, a physical form would not be needed, just the computational capability to do the mathematical operations and relaying of messages needed for commanding the physical robots. For example, the C2 agents may also be coordinating robots with human relief personnel so that the robots with the needed capabilities are where they are most needed and the humans are filling those roles that only a human could. Logistics agents would also be performing functions to make sure equipment, material, food, and other needed items are where they need to be at the right time. There may be cases when the robots are acting as assistants to the humans and other cases when the humans are acting as assistants to the robots, based on who has the most expertise, skill, or strength given the needs of the situation.

Such a collection of robots and humans could act as an effective disaster relief team. The robots could perform the dangerous and heavy lifting tasks as well as tasks such as setting up tents, moving supplies, and searching for survivors. Command and control agents could be collecting information from the robots as they survey areas, find people, discover people with medical needs, and perform forward supply chain functions.

As an example, the robots may find that a road between towns is blocked and needs to be cleared or another way found to the town. Memes would have to be searched for ways to clear different types of debris, perform simulations to determine if the memes will work and then modify them, or ask for help. It might happen that the local robots do not have the ability to move the debris (it is too heavy, large, etc.) so they need to find a new path to the town. The robots update their memes to include that the road to the next town is blocked and impassable and then share it with the command and control robots/agents and others.

The robots and supporting agents then begin searching for a new path to the town. Other paths are explored on maps, through the community of robots and agents, and then physically to make sure that the potential new path is not blocked. This path then forms the basis of a meme that describes the path and is shared among the robots in the community. The meme would need to have all of the information described above in Eq. (4.1) for a CMC_R. It would need a memetic engineer (creating agent) to construct it, a hook and bait (problem and results) to get other robots' attention, a medium to share the meme, a host or sending agent, information on the new path, and a description of what environment/community the meme is applicable to.

An example of a CMC_R for the path from the first town to the neighboring town would be the following:

CMC_R-path $=$

$\qquad$ *meme name* $=<$ *town path* $>$,

$\qquad$ *creating agent* $=<$ *HADR-Agent*1 $>$,

$\qquad$ *problem* $=<$ *Path to Town*2 $>$,

$\qquad$ *results* $=<$ *Way Points* $>$,

$\qquad$ *network* $=<$ *HADRnet* $>$,

$\qquad$ *sending agent* $=<$ *HADR-Robot*3 $>$,

$\qquad$ *meme* $=<$ *waypoint*1, *waypoint*2, *waypoint*3, *waypoint*4 $>$,

$\qquad$ *community* $=<$ *HADR-Group*1 $>$ $\qquad\qquad\qquad$ (4.2)

The above meme has HADR-Agent1 as the engineer of the meme, "Path to Town 2" as the problem that is solved (hook), "WayPoints" as the solution to the problem (bait), "HADRnet" as the transmitting network, HADR-Robot3 as the host that sent the meme to the receiving robot/agent, the waypoints as the memotype/meme content, and HADR-Group1 as the agent and robot community in which it was generated. From this meme, the newly discovered path between the towns can be shared among the robots and other systems in the relief effort. Each robot would then save the meme along with other memes that would form a "Meme Knowledgebase". When the robots from Town 1 reach Town 2, the meme can be shared with the new community of robots in Town 2, which in turn would store it in their meme knowledge bases.

In our example, after the new path meme is created and shared, other agents use the meme to re-evaluate their current memes, updating their memes in an evolutionary fashion by replacing, modifying, or combining their current memes. In this example, after the logistics robots receive the updated path meme, they then update their other memes, one of which might be a new command location because of the new path to Town 2. With the new path information to Town 2, the logistics robot performs analysis and decides to move supplies to be in a more optimal position between towns and updates other operational constraints. From this, a new meme is engineered and distributed to the other robots/agents, and the process propagates through the community.

4.8   Conclusion

This chapter introduced ideas for what is required in order for robots to perform memetic algorithms. Included are means and mechanisms for meme observation, self-awareness, evaluation and manipulation. It also introduced a formalization for robot memetics based on the formalism of an intelligent meme. Examples leveraging the formalism were presented in a scenario set within a disaster relief context in an uncertain environment where memetic algorithms would be used.

The next chapter covers examples of the above ideas on robot memes and memetic algorithms suggesting the manner in which robots can acquire intelligence, exchange ideas, and cooperate with one another and with humans to accomplish goals. The examples involve robots learning new skills through observation, self-evaluation, formalization of memes, and problem solving situations that suggest how robots, with help from humans, can become aware of their environments, cooperate and collaborate, create a culture, and gain new intelligence that can be shared with other robots and with people. The backdrop of a future Mars settlement provides a rich set of possibilities to use the above ideas in a range of scenarios that can hopefully motivate new scenarios and supporting technologies to develop self-aware robots.

References

1. Le M, Neri F, Ong Y (2015) Memetic algorithms. computational intelligence, vol II. Eolss Publishers, UK, pp 57–86
2. Huntsberger T, Stoica A (April 2010) Envisioning cognitive robots for future space exploration. In: Proceedings of SPIE multisensor, multisource information fusion: architectures, algorithms, and applications, vol 7710, 77100D-5, Orlando, FL
3. Eberbach E (2005) $-Calculus of bounded rational agents: flexible optimization as search under bounded resources in interactive systems. Fundam Inform 68:47–102
4. Wang Y (2015) Concept algebra: a denotational mathematics for formal knowledge representation and cognitive robot learning. J Adv Math Appl 4:1–26

Chapter 5
Mars Settlement Scenario

5.1 Introduction

Planetary surface exploration is a pursuit of interest for countries around the world. Robots and robotics technology play a critical role in exploration today, and will increasingly in the future when robots will be very intelligent and autonomic. For such robots, the range of activities associated with planetary exploration and settlement present a worthy canvas for illustrating the potential of robot memetics.

Consider the existence of a human and robot settlement on Mars, the backstory for which holds that the international Mars exploration campaign of today led to discoveries of habitable regions with accessible water and other in situ resources conducive to human settlement and sustenance by living off of the land. Viable approaches are developed for dealing with the quite formidable obstacles keeping the surface of Mars from being hospitable to humans (lack of accessible water and oxygen, unsafe surface temperatures, toxic soil, and ultraviolet radiation). Multiple sites are characterized by progressive robotic science missions and established as candidates for settlement. The candidate sites include regions at the surface and near subsurface where habitable (protected) structures, to which water and oxygen can be delivered, can be installed by robots. Among them, a site dubbed Plannex-A was chosen for a first settlement. A campaign of missions, as precursors to human presence, sent robots and equipment to the site to prepare it by constructing an outpost with habitats for humans, power plants, laboratories, medical facilities, mines, etc., and to further emplace and install infrastructure, as is illustrated conceptually in Fig. 5.1.

The first humans on Mars are yet to arrive, but have been rehearsing their future experience sequestered in analog settlements on Earth in cohabitation with intelligent robots. Thus, the social human–robot interaction dynamic is already established and evolving in preparation for it to play out on the Martian surface when humans arrive at Plannex-A with additional robots to integrate with the robots already there (those that have prepared and are maintaining the settlement site).

© The Author(s), under exclusive licence to Springer Nature Switzerland AG 2020
W. Truszkowski et al., *Robot Memetics*,
SpringerBriefs in Electrical and Computer Engineering,
https://doi.org/10.1007/978-3-030-37952-0_5

Fig. 5.1 Conceptual Plannex-A settlement during establishment *Courtesy* NASA

Meanwhile, robots inhabiting Plannex-A are performing daily duties of settlement maintenance while interacting with each other and occasionally with remote humans (on Earth or in a crewed space station or spacecraft temporarily orbiting Mars). This is the backdrop facilitating the interaction fostering robot memetics and early emergence of a human–robot society at Plannex-A.

5.2 Mission Outline

The overall mission at Plannex-A is for humans to establish and maintain a research colony with support from heterogeneous robots with different capabilities whose initial function of establishing the physical settlement and preparing it for human presence has been fulfilled. In support of the overall mission, these memetic robots serve the functions of settlement maintenance and sustainment as collaborative partners to humans. Human crews come from Earth and return to Earth over the course of conducting a series of expeditions, leaving robots for periods of time as sole inhabitants of the settlement. As a broader goal, the aim is to mature the settlement from its primary function as a research outpost to a permanent human–robot colony on Mars.

5.3 Robot Community

The vitality of Plannex-A relies on a cast of memetic robots of different categories capable of performing different tasks to accomplish goals of utility to the settlement and human endeavors within. Included is a corps of construction, exploratory, and research robots with manager robots as coordinators. These robots must interact and work together, as well as with humans to accomplish the mission. Sharing memes, modifying memes, and creating new memes will allow them to work in this unstructured environment by sharing information, creating a community with a culture, and changing that culture to persistently accomplish the mission goals. This corps of memetic robots that might work together on Mars is depicted in Figs. 5.2, 5.3, and 5.4.

a. Cast of Robots

Construction robots, designated as the C-series, build infrastructure both for robots currently on Mars and for future human habitation. *Exploratory robots*, designated as the E-series, find new areas of the settlement region to explore and prospect for in situ resources. *Research robots*, or R-series, analyze atmosphere, soil, and geology as well as perform computational analysis in support of research. *Manager robots*, M-series, are humanoid in form and coordinate the activities of the various robots so that they do not interfere with each other and so that tasks are prioritized (e.g.,

Fig. 5.2 Conceptual depiction of humans and heterogeneous memetic robots working together at the Plannex-A Mars settlement

Fig. 5.3 Cast of robots (pictured left to right: C-series, E-series, R-series, M-series) based on appearance/adaptations of existing and prototype NASA robot designs (the ATHLETE, K10, and Curiosity planetary rovers, and R5 Valkyrie humanoid robot)

Fig. 5.4 C-series robot featuring attachments for construction (top left), E-series robot featuring high mobility and lightweight features (top right), R-series robot featuring science instruments and analysis tools with more sophisticated processing (bottom left), and M-series robot featuring humanoid form (bottom right)

construction robots build shelters before building roads, or research robots perform research in areas explored by exploratory robots). M-series robots also exploit their anthropometry to perform many of the same tasks that humans can, including tasks calling for the use of human tools as well as facilities and vehicles designed for human operation.

Additionally, the robot population includes medical robots (for human health-care), mechanic robots (for robotic and non-robotic equipment), and transportation robots. As such, Plannex-A is a multi-species environment involving human–robot and robot–robot interactions. For the latter, the diverse cast of robots and potential for memetic evolution can permit the emergence of different additional species of robots. Similar ideas have been envisioned for nearer-term plans for outposts on Earth's moon, such as the Moon Direct mission concept [1] wherein various rovers would be teleoperated to provide different task-based utilities. For Moon Direct, some rovers would support the setup of solar power and communications systems, others would serve as scouts to perform landing area surveys, and other rovers would distribute and precisely emplace radio beacons to support future landings.

b. Preloaded Memes

For all robots in the scenario, initial memes are preloaded by humans before Earth departure as core competencies sufficient for reliable execution of mission tasks and for survivability in the unstructured Mars surface environment. That is, humans provide initial (bootstrap) memes to seed robotic intelligence based on what humans want the robots to achieve and what they think the robots will encounter on the mission in terms of work to do and challenges to overcome.

The basic set of preloaded memes representing core competency sets can differ for each category of memetic robots, enabling each type of robot to perform its intended function(s). Examples of core competencies for the construction robots include how to set up structures on Mars in particular weather conditions and on particular terrain at settlement locations as well as what the relative emplacement should be. For exploratory robots, core competencies include how to navigate terrain in the region of Plannex-A, where exploration should start, guidance on where to find resources in the region, and how to use an initial set of tools and instruments on the robots to dig for, extract, or measure resources of interest. For research robots, core competencies include data gathering, sample acquisition, and analysis techniques for atmosphere, soil, and geology based on research questions and associated experiment designs of human scientists. Manager robots are endowed with task planning, scheduling, and execution monitoring capabilities as core competencies along with group manage-ment techniques for a set of robots and anticipated modes of human participation. In addition, they are seeded with skills for performing a range of useful physical tasks with human or superhuman capabilities using tools, equipment, and systems designed for human operation.

Memetic robots in the broader population would also have preloaded core com-petencies. For example, medical and mechanic robots will know how to treat various human and robotic equipment ailments based on embedded expert knowledge and experience from earlier robotic Mars missions as well as cohabitation experiences in analog settlements on Earth. All of these core competencies will be expected to be expanded over time based on what is encountered on Mars and on what is needed to accomplish mission goals. As unforeseen circumstances arise, new memes will need to be created, or existing ones modified to address those circumstances thus enabling adaptation. Examples of this are played out later in the scenario as the result of robot memetics.

Starting with core competency memes, robots progress through their experiences on Mars applying the memes to accomplish initial goals. Through meme transfer and memetic learning, the robots immediately start updating their meme knowledge base with information on their surroundings, via observations of other robots, and via interactions with humans. That is, by interacting with other robots, the environment, and humans, memetic robots at Plannex-A are forced to evolve and adapt as needed (due to selective pressure, in a Darwinian evolution sense).

c. Memes and Community Development

Just as memes can change the way people and robots think and act, memes can also physically change a community and its culture, whether made up of people or robots, or both. As described in Chap. 2, as new memes are learned, new ideas can be formed by linking the memes together in new ways to form complex memes, modifying some memes to make them fit together in a complex logical structure and generating new memes to fill in missing gaps. New ideas then lead to new actions and those actions can change not only how robots or people interact with each other, but also a community's culture and its physical surroundings.

Memes can change a community physically as well as its culture through new ideas on ways to work—which can lead to where a community lives, and how structures and buildings are built based on the new location and work. The new work, based on the new memes and ideas, may require new tools, processes, goals, communications, and organizational structure, which can change the community physically. On Mars, this could include a new meme expressing that the current location is more susceptible to high or abrasive winds that cause structural damage to settlement infrastructure. Moving the structures to a more sheltered location would reduce the amount of time to repair the settlement from storm damage and provide more time to be spent for exploration and research. More subtle physical changes could also occur. It could be determined that instead of needing to move the whole community, only certain parts of the structures in the settlement would need to be made stronger and would be able to withstand damage from wind-blown dust or to better protect against radiation.

Memes can also influence the splitting of a community. For example, given that robots on Mars would not need life support systems as humans would, a robot community might evolve its culture in such a way that drives changes to its physical environment that make it more efficient for robots to perform work without life support systems nearby. This could manifest during periods when only robots are present at the settlement and could result in humans, upon return to the settlement, finding a need to travel longer distances to and from a worksite, between areas where there is life support and where the robots tend to work. A compromise may need to be reached between the robot and human factions of the community where an area exists at or near the worksite that contains life support systems in addition to a larger life support area where humans would live and work. Memes would have to be generated that include efficiencies and tradeoffs between the communities to provide overall human–robot colony efficiencies.

Linking memes and changing memes into ideas offers the possibility to make both large and subtle changes to a community and its culture. If problem solving is not converging, or viewed as going in the wrong direction, memes can be added to the community to nudge problem solving in a new direction. We see this, for example, in human groups that have a facilitator helping a brainstorming or other group avoid going off track as it works on a problem. The facilitator can make statements (memes) that coach the group back to solving the correct problem within the given bounds. This can also happen with a human, or other robot or an agent, that interjects memes to keep robots from emplacing structures too far apart, exploring in unwanted areas, or performing other unwanted actions. These memes can come from a hierarchy of sources, perhaps first from a local supervisory robot, then from a higher-level supervisory agent, and then lastly from humans in charge of settlement construction/operations and the overall mission.

5.4 Robotic Activities Ongoing at Plannex-A

In preparation for human arrival at Plannex-A, E-series robots are prospecting for in situ resources needed for increasing the supply of potable water. Their exploration is guided by specifications and instructions on where in the Plannex-A region they might find critical resources from which water could be extracted. These specifications and instructions were given through a set of memes during the start of the mission. The information and instructions on the locations of critical resources is not complete but is sufficient for E-series robots to execute the exploration task with incomplete or even partially correct information through the use of exploratory behaviors and memetic learning techniques. For instance, the memes provided to the E-series robots suggest that in order to find a specific resource they need to start searching and/or excavating in specific areas of Plannex-A and its outskirts, and their memes may be seeded with coarse routes in the form of waypoints separated by substantial distances. The C-series robots have the responsibility of ensuring the readiness of infrastructure for processing in situ resources and storing their important byproducts, and also have memes expressing the locations best suited for storage and safekeeping. The R-series robots are initialized with memes that instruct them on how to analyze and assess the resources gathered by the E-series robots. Such memes enable them to help E-series robots determine whether or not they are gathering resources of sufficient quality and quantity. M-series robots are coordinating and monitoring the progress of overall robotic activities.

Such is representative of days in the lives of memetic robots at Plannex-A. Impacts of robot memes and memetics are illustrated next in the contexts of several example scenarios wherein challenges are encountered at the settlement.

5.5 Challenges, Unexpected Problems, and Memetic Solutions

a. Dust storm

On a given day, a dust storm impedes progress of robotic activities and degrades communications throughout a portion of the Plannex-A settlement. Because of the severity of the dust storm, visibility throughout the environment is reduced to the point where it is no longer safe for the robots to travel, so they have to stop. The robots are unable to get to a shelter, so they remain exposed to the blowing dust. The result is that some of the robots are damaged. When the dust storm stops, M22, an M-series robot, notifies all robots/agents that the storm is over. An R-series robot additionally broadcasts a new capability to forecast dust storms, a result of collective data gathering by R-series robots during prior wind-blown dust events and analysis by remote humans. From this most recent experience and new knowledge shared by R-series robots, M22 also learns a new meme suggesting that the robots should seek shelter as soon as possible whenever there is a possible dust storm according to some forecast. The new meme in this case enhances safety of the robot community through new knowledge shared by M22. Damages sustained may also inform the mission team back on Earth regarding the manner in which future robots sent to Mars would need to be better protected from blowing dust, and may thus influence the physical design of future robots that will join the settlement.

After the dust storm, undamaged E-series robots continue to explore their environment and one of them, E11, discovers that a path that it has planned to travel to reach a resource location is blocked. E11 then updates its memes for following routes to that location with information that its planned path is blocked. This information is shared with the other robots and agents, who in turn share it with other robots and agents so that the whole community eventually becomes informed. For some agents, this information will be of little use since they are not concerned with paths to resource locations. After sharing their updated memes, E11 and other exploratory robots affected begin searching for any new paths to the same or other resource locations.

To generate new paths, new routes between the resource locations are explored using exploratory behaviors along with any prior and newly acquired knowledge of the environment until new paths are discovered. A newly found path then forms the basis of a meme that describes the path and is shared among the robots in the community. The meme would need to have all of the information described in Chap. 4, Eq. (4.1) as comprising a CMC_A. That is, it would need a memetic engineer to construct the meme, a hook and bait to get other robots' attention, a medium to share the meme, a host or sending agent, information on the new path, and a description of what environment/community the meme is applicable within.

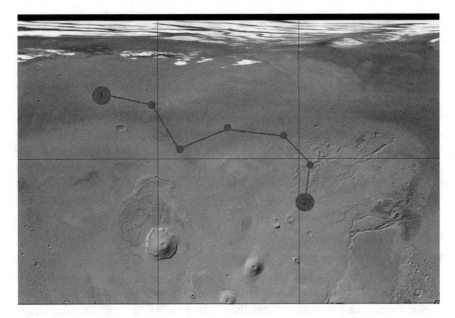

Fig. 5.5 Map illustration of CMC$_A$ for the path from Resource Location 1 to Resource Location 2

An example of a CMC$_A$ for the path from Resource Location 1 to Resource Location 2 is the following (Fig. 5.5):

CMC$_A$-path =

        meme name =< *resource path* >,

        creating agent =< *E - Robot*1 >,

        problem =< *Path to Resource Location* 2 >,

        results =< *WayPoints* >,

        network =< *MarsNet* >,

        sending agent =< *E - Robot*3 >,

        meme =< *waypoint 1, waypoint* 2, *waypoint* 3, *waypoint* 4, *waypoint* 5>,

        community =< *Plannex - A* >   (5.1)

The above meme has exploratory robot 1 (E-Robot1) as the meme engineer, "Path to Resource Location 2" as the problem that is solved (hook), "WayPoints" as the solution to the problem (bait), "MarsNet" as the communications medium for meme transmission, E-Robot3 as the E-series robot host that sent the meme to the receiving robot, the waypoints for the resource path as the memotype/meme content, and Plannex-A as the agent community in which it was generated. From this meme, the newly discovered path between the locations can be shared among the robots and other systems of Plannex-A. Each robot would then save the meme along with other memes that would form their evolving "Meme Knowledgebase."

In this example, after the new path meme is created and shared, other robots use the meme to re-evaluate other memes that they have, updating them in an evolutionary fashion by replacing, updating, or combining their current memes. After an E-series robot, in this case, receives the updated path information, it updates other memes— one being the feasible path to the resource location. Also, existing memes may be impacted calling for updates based on the updated path meme, such as those related to a charging station optimization scenario discussed in more detail later. In that case, an updated meme is formed based on a combination of multiple memes and is thus a "complex meme," as discussed in Chap. 2. In a similar manner, new memes are engineered and distributed to the other robots/agents, and the process propagates through the community.

b. Solar Array Disruption

The power generation infrastructure at Plannex-A includes solar arrays assembled by the memetic robots using core competency memes. C-series robots, in particular, contributed using memes conveying solar array construction plans as well as manipulation memes for handling and installing solar panels that comprise solar arrays. On another bad weather day at Plannex-A, a dust storm is forecasted and shared with robots by the settlement's weather service. Thanks to the recently learned and shared meme for seeking shelter from impending dust storms, robot storm damage throughout Plannex-A is minimal after the storm passes. However, M12, an M-series robot conducting a post-storm damage assessment reports damage to a panel in the solar array and a need for its replacement. M12 tasks C-series robots, C22 and C30, with fetching a replacement solar panel and replacing the damaged solar panel. M12 and other M-series robots have memes enabling them to use tools and equipment designed for humans. M12 uses such tools and relevant memes to disengage the damaged solar panel and to transport it to a repair facility using a human-drivable, unpressurized transportation rover while C22 and C30 retrieve and cooperatively transport the replacement solar panel from a spares storage facility to the solar array.

C22 further observes an additional problem—a foundation leg, or pile, for the solar array is also structurally compromised due to surrounding dust storm-induced erosion, and at least one new hole needs to be drilled to re-establish a firm foundation for the solar array structure. To ensure an informed selection of a new location for the foundation leg, a human assesses the situation based on data and imagery acquired by C22 and available in the Plannex-A cloud, and decides to dispatch an R-series robot to sample various patches of nearby soil to generate additional data supporting the decision. R2 is available and arrives on the scene to acquire ground-penetrating radar and penetrometer measurements using a meme acquired from observation of human field geologists, but adapted through memetic learning to enhance performance and accuracy accounting for the R2's capabilities relative to those of a human using functionally similar, but different tools. R2's investigation reveals the local subsurface structure that shows a variable stiffness distribution allowing localization of a suitable area for supporting the foundation leg based on maximum stiffness. This provides the desired recommendation for the drill-hole location.

A meme associated with solar array construction prescribes drill-hole depths of at least 0.5 m to be adequate for solar array supports based on prior Earth-based experiments and prior knowledge of geology in the region. Accordingly, C22 proceeds to drill a 0.5 meter hole at the R2-recommended location. While drilling the new hole, measurements from C22's drill sensors indicate unfavorable soil properties (for sufficient structural support) at approximately 0.5 meters depth. C22 continues drilling to sample the conditions at just beyond 0.5 meters, only to be met with unexpected readings contradicting the hole-drilling knowledge prescribed by its solar array construction memes. As an autonomic response to erratic vibration and intermittent drill-bit stiction, C22 retracts the bit and changes to a different bit and drilling technique to try a meme learned from past experience when drilling under similar conditions and sensed effects. Rather than abort the task, and having changed the drill bit and technique, C22 decides to resume, drilling further to find favorable soil conditions at a depth of 0.8 meters and beyond. C22's inquisitive task exploration results in new knowledge suggesting that drill-hole depth at Plannex-A should be at least 1 m. C22 then uses its original memes for solar array construction with this new knowledge of minimal hole depth to proceed with restoring structural integrity of the solar array.

During the experience, C22 exhibits several of its autonomic attributes and properties for self-management including environmental- and self-awareness, self-monitoring, self-adjustment, and self-configurability. The new knowledge it gains is shared with other C-series robots and with humans in mission control on Earth, prompting formulation of new experiments for R-series robots to conduct that would verify the consistency of the new knowledge and improve the understanding of geology at Plannex-A. Later, R-series robots would team to conduct geological and soil property surveys throughout the broader terrain surface with consideration of the new knowledge that C22 acquired regarding minimum drill depths of 1 m. Meanwhile, C22 and C30 proceed with re-establishing a firm solar array foundation and installing the replacement solar panel.

c. Charging Station Optimization

Solar arrays are one source of energy feeding energy storage systems at the Plannex-A settlement, some of which are connected to charging stations for robots. In the beginning of the settlement's establishment, there were a limited number of charging stations for the robots. Smaller robots like the E-series need about one hour to fully charge while larger robots like the C-series require up to four hours to function for a full day. The limited number of chargers throughout the settlement cannot support the entire population if more than one-third of the robots need to charge simultaneously. Crowding at charging stations has become a problem as robots are more commonly unable to meet their mission schedules because they are forced to wait in queues at inopportune times to charge their batteries. Recognizing that the cause of their scheduling setbacks is due to the time spent waiting for an available charging station, the robots decide to develop a solution to share a limited resource among themselves in a way that most efficiently benefits the society. Specifically, M-series robots are tasked with the challenge of developing a new meme to solve this problem.

The M-series robots do not yet have a meme specific to solving this issue in the meme pool and therefore must develop an original solution. Manager robot M102 identifies the problem as an issue related to a lack of scheduling organization and prioritization of the charging stations. M102 searches through the memotypes in the meme pool to find memes related to scheduling optimization and prioritization and comes across a variety of prioritization memes for different applications ranging from triage instructions after a catastrophe to priority scheduling for ongoing and future mission tasks. In addition, M102 evaluates memes generated based on battery charging and discharging rates for each species of robot. These memes were selected based on memotype attributes that suggested relevance to the charging station issue. M102 processes the relevant memes and develops multiple proposed solutions according to the complex memes formed by combining concepts in the relevant memes. Based on its evaluation, M102 derives a charging priority schedule for the population based on the mission schedules and species-specific battery requirements. The proposed charging schedule attempts to optimize the amount of time spent at a charging station with the importance of the individual robot's mission schedule and its projected battery use. Additionally, M102 observes that the time robots spend at charging stations would reach peak efficiency if there were 20% more charging stations in the settlement.

The proposed solution is implemented in the community and the settlement does experience lower average amounts of time spent by robots at charging stations. However, there are unforeseen consequences of the implemented schedule. Since the new charging-schedule meme prioritizes charging according to mission task priority, less important tasks have been affected by longer delays than expected. Some of the lower prioritized mission tasks indirectly affect the success of the more important tasks. Delays in the low priority tasks have resulted in longer overall delays for the community's mission schedule. Robots working on less important tasks run low on battery charge more often and have higher average charge times. M102 recognizes that the charging-schedule meme needs to be further tweaked because of the drawbacks to overall mission efficiency. Using the memes learned from the initial trial, M102 performs simulations to rebalance how priority is assigned to different robots based on their species and mission schedule.

The information learned from the initial trials is also analyzed by other manager robots. In particular, another manager robot, M200, processed the newly learned memes and suggests adding additional charging stations to further alleviate the problem. M200 also performs a series of simulations and finds that if the charging stations were dynamically relocated based on mission schedules and topography, then the charging-schedule meme would be even more efficient. M200 provides instructions to the C-series robots to physically relocate the charging stations. The C-series robots perform the relocation according to schedules and locations provided by the memes generated by the M-series robots. An example of a CMC expression of a relocation meme processed by a C-series robot would be the following:

$CMC_C =$

$$meme\ name\ =<ch\arg ing\ station>,$$
$$creating\ agent\ =<M200>,$$
$$problem\ =<Ch\arg ing\ Station\ 5\ Location>,$$
$$results\ =<Time\ and\ New\ Location>,$$
$$network\ =<MarsNet>,$$
$$sending\ agent\ =<C101>,$$
$$meme\ =<Time\ to\ perform\ relocation,\ Coordinates\ of\ current\ station$$
$$location,\ Coordinates\ of\ proposed\ station\ location>,$$
$$community\ =<Plannex-A>\qquad\qquad (5.2)$$

After many simulations and trials of implementing different charging schedules, the settlement adopts a scheduling meme that solves the problem of crowded charging stations. Also, based on the recommendation made by M200, there are plans to add new charging stations to further alleviate the problem and also relocate charging stations based on future terrain and mission schedules. Each robot now receives a notification of what time they are to report to a specified charging station in order to optimize the overall use of the limited number of charging stations while staying on track for mission schedules. An example representation of the final developed battery charging meme can be visualized as in Fig. 5.6.

d. Disabled Robot

With a settlement now supporting more reliable and efficient sustenance for robot power, exploratory and research robots become more free-ranging and can afford to explore in numbers and over greater distances. E-series robots E2, E105, and research robot R5 are navigating through an unexplored area while searching for

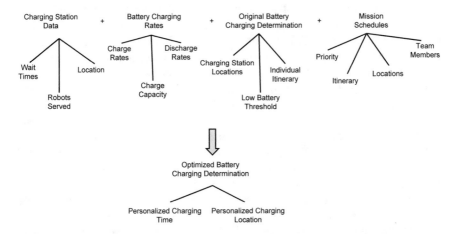

Fig. 5.6 Mission-related memes combined to form optimized battery charging determination

new places to prospect and mine for water resources. The current global map of the area indicates that all three robots should be able to navigate through this area without any traversability or trafficability problems. While negotiating the local terrain, E2 gets stuck due to a large hole covered by loose soil and not detected by its onboard sensors as a mobility hazard. E2 was an early version of the exploratory robots and has smaller wheels and is not able to traverse as rough a terrain as E105. E2 uses its proven memes as well as some newly updated memes to try to extract itself from the hole. After a half-hour of trying, it comes to the conclusion that it is stuck and notifies E105 and R5 of its circumstance.

E105 comes to the area where E2 has become stuck while R5 continues to perform some research in the explored area. E105 takes images of the area, passes them to E2 to help it find a solution to getting unstuck. After analyzing the images, both E2 and E105 determine that E2 has already taken the steps most likely to be effective at extracting itself from the hole and that trying different approaches will likely fail and only result in further reduction of its battery charge to risky levels.

E105 has a meme for extracting other exploratory robots that have gotten stuck in loose soil or hung up on rocks, which are the two most common ways robots have gotten stuck in the past. The meme calls for E105 attaching its arm to a tie-down attachment on another exploratory robot and both robots moving at the same time while attached. With the extra pull of the second robot, a stuck robot usually becomes free. E2 is an older heavier exploratory robot, so the attempt fails in this case.

At this point E105 sends information back to the manager robot M3 to let it know that E2 is stuck in a hole, that E105 has used a current meme that did not work, and asks if there are any other memes in the meme pool that might be applicable. M3 examines the data from E105 and performs simulations on the amount of force that would be needed to get E2 out of the hole given images of the hole and soil information. After running some simulations and some genetic algorithms based on the current meme, M3 determines that the combination of E2 and a construction robot should produce enough force to pull E2 out of the hole.

The nearest construction robot is contacted, but it is involved in constructing a habitat for humans, and that project is behind schedule. Taking the construction robot off of that project would further delay it; and since humans are on their way, it is imperative that it is finished on time so that their pressurized quarters are ready prior to landing. The next nearest construction robot is several hours away and would need to be recharged before making the trip, meaning it could be at least 4 h before it arrives, placing its estimated arrival time close to nightfall.

In the interim, M3 also sends a request for help to the human engineering team located on Earth. A picture of the stuck robot, a summary of the problem, diagnostics of the robots involved, and the current attempted and proposed solution is forwarded to the Earth-based team for guidance. The human engineers brainstorm ideas and suggest to M3 to also attach R5 to E2 to generate greater pulling force to extract E2 from the hole. In addition, the human engineers give guidance about the best angles to place the robots relative to each other to pull E2 out of the hole.

Given this information, the manager robot, M3, runs the meme through more genetic algorithms and does more simulations and determines that the joint force

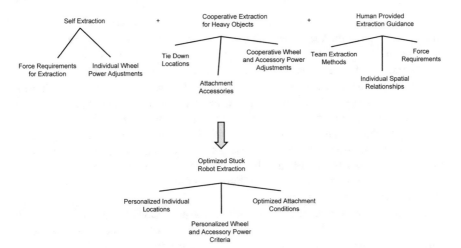

Fig. 5.7 Extraction memes combined to form optimized stuck robot extraction meme

from the robots E105 and R5 applied to E2 should meet the minimal amount of force needed to pull E2 out of the hole. This information is sent to E105 and R5, and R5 joins E105 and E2. R5 is able to attach an arm to another tie-down attachment on E2, and with E105 and R5 pulling on E2, they are able to extract E2 from the hole.

The cooperation between robots and human engineers resulted in new knowledge learned by both communities. Aspects of the newly learned meme can be visualized as shown in Fig. 5.7.

Several memes are updated based on the above experience and its trials of failure and success. One expresses that the older exploratory robots like E2 cannot get out of a hole of that size. A second meme expresses that the path through that area is not safe for older exploratory robots. A third meme expresses that a newer exploratory robot and a research robot can pull an older exploratory robot out of a hole, not just out of loose soil or jagged rock formations. So, there are now several new memes emerging from this experience.

e. Human–Robot Symbiosis

During the early months since human astronaut arrival at Plannex-A, effects of radiation overexposure during space travel plus additional long-term exposure on Mars befall an astronaut in the form of a chronic radiation condition. Astronaut "Conrad" exhibits cognitive confusion, impaired vision, and an apparent reduction in curiosity—all symptoms experimentally observed to result from long-term radiation exposure in mice. In recent weeks, Conrad's duties found her spending increasing amounts of time working and interacting with one of the humanoid M-series robots, M3. This period of inflection, wherein physical human presence has been suddenly increased relative to the robots' experience thus far at Plannex-A, sees an increase in robot observances of human behavior and work practices. By now, M3 has benefited from high meme activity induced by hours spent with Conrad, and the same is true for other M-series robots by virtue of their anthropomorphic similarity to humans. M3 has acquired a number of effective memes, increasing its knowledge and capability as an M-series robot. However, Conrad's chronic radiation condition has set the stage in recent weeks for M3 to have adopted or learned some "infected memes" in addition to effective ones. A case of bad memes as discussed in Chap. 3 has arisen.

As astronaut Conrad recovers, M3 eventually determines via autonomic self-awareness that its performance has degraded and its behavior is uncharacteristically ineffective. Not to mention, one or more of M3's infected memes have spread to other robots. Informed by this self-observation, M3 can now detect the infected meme(s) in other affected robots. Through further observations of healthy humans, interactions with them, and requested human assistance, M3 acquires memes that ultimately facilitate its self-repair.

With M3 once again considered cognitively astute, and with other robots able to assess the utility and effect of M3's new memes, bouts of high meme activity occur during which M3 shares its memes via the meme pool and other robots observe M3. This ultimately results in the "healing" of other affected robots in the community. In addition to having acquired effective memes from humans and memes to counteract infected memes derived from degraded human behavior, M3 acquired a meme that could be particularly important for the future of the human–robot community at Plannex-A—"trust, but verify"—a meme for second-guessing human-derived memes.

Experiences such as M3's and those of other very intelligent robots interacting with humans continue toward accelerating forms of human–robot symbiosis. In this case, astronaut Conrad's experience with chronic radiation reveals to the mission personnel that its instruments and other means to monitor, detect, and diagnose the condition in the most timely fashion are not sufficiently effective at Plannex-A and perhaps on Mars at large. Over time, M-series robots develop memetic algorithms that go beyond detecting infected memes in other robots to facilitating detection of chronic radiation in humans by observation of their behavior. As such, robots become the primary means to do so while the bio-instrumentation human-engineered for Mars becomes less relied upon for that purpose.

5.6 An Evolved Scenario

At this point in the scenario, it has been two years since memetic robots have landed on Mars and started developing a robot community and a habitat for humans. Humans joined the robots a year ago and they are working to evolve an ever-improving community where both humans and robots can work and explore together (Fig. 5.8). The robots have learned the needs of humans, their work/eat/rest/play schedule and have adapted their behavior to remain energy-positive via self-charging, to perform maintenance, and more, so that they are ready to work when humans are ready and maximally productive otherwise. The robots have also learned what the humans are good at and not good at. They can now rely on humans to do what they do best, having

Fig. 5.8 Depiction of human–robot presence at the Plannex-A Mars settlement

learned to support the humans where they have their biggest challenges. For example, the robots learned that the humans did not like being outside during even mild storms, and preferred not to work after nightfall. With this knowledge, the robots performed charging and maintenance during such times, and performed robotic exploration and research tasks on schedules ensuring the results would be ready for the humans when they started working again. Whenever humans returned to Earth or orbiting space stations, robots at Plannex-A would fall back to schedules and workflows efficient for the solely robotic community.

The humans in turn learned the robots' needs of charging, maintenance and work, and have adjusted their own schedule based on the robots' needs. Humans have also learned the strengths and weaknesses of robots and know when to use which robot for a task versus when a human is best to perform it. Humans have also learned how to best introduce new memes into the meme pool to help robots solve a problem, or to modify robots' behavior to help them better assist the humans. For example, humans have introduced new memes to direct robots toward new research based on human analysis as well as new memes to get help from the robots on the more mundane or dangerous tasks the humans were performing that would be best done by a robot. Humans have also introduced new memes that would directly allocate particular tasks that humans would perform better to humans, such as drawing conclusions from research and conceiving the progressive steps to take toward permanent settlement on Mars.

This learning and adaptation by both robots and humans has created a new human–robot culture. The human–robot interactions have become almost subconscious for the humans as the humans and robots start to work together as long-time partners. The robots in turn have developed a fairly stable set of memes that allow them to work seamlessly with the humans. Humans have made explicit introduction of new memes to directly influence the human–robot community, and robots have added memes based on observation of human activity, as well as through their normal learning and day to day experiences on Mars. As new humans and robots join the community, they have to learn how to interact in the established culture, but the introduction of each new human and robot contributes new memes to the joint human–robot culture and it evolves to create an ever-increasing partnership in moving the community toward its goal of understanding and settling on Mars.

Reference

1. Zubrin RM (2019) Moon Direct: a cost-effective plan to enable lunar exploration and development. AIAA Scitech 2019 Forum, Paper# AIAA 2019-061, Jan 2019, San Diego, CA

Chapter 6
Conclusion

A shared society between humans and robots is already beginning to form on Earth. The emergence of autonomous passenger vehicles, service robots, service agents, and household agents is quickly resulting in a more integrated human–robot community where humans and robots are beginning to develop a symbiotic relationship. Yesterday's nascent mobile automata and early adaptive machines, along with today's planetary rovers and humanoid robot partners, are laying a foundation for a not-too-distant future in which robot memetics will facilitate advanced robotic intelligence and future human–robot collaboration.

In that future, robots capable of complex learning will exist alongside humans. Memetics can form the framework for this learning and information sharing between robots and humans in the future shared society. As knowledge evolves, so too does culture. Human societies have been able to record cultural information and evolve cultural practices over generations based on community knowledge. Robot communities will similarly be able to develop a knowledge base that will serve as a foundation for cultural evolution and propagation through the use of robot memes.

Memes can serve as a means for not only storing information, but for active information exchange as they can be self-modified via intelligent memes or manipulated by members of the society. Robot memes can be formally represented using the model presented in Chap. 4 which enables knowledge to be readily distributed and analyzed. Next-level robotic intelligence can be developed using formal models for robot memes along with concepts of intelligent memes and memetic algorithms. In addition, an associated framework lends itself to robot meme interpretability and explanation, which can facilitate human understanding of intelligent robot behavior, whether programmed, learned, evolved, or emergent.

A knowledge base with a memetic foundation will enable communities to thrive and evolve in dynamic environments where adaptation is an essential trait for survival. This principle is illustrated in Chap. 5 and exemplifies how a shared human–robot society on Mars could evolve to perform its mission given an uncertain and/or unknown operating environment. Underlying that societal evolution are robot memes

© The Author(s), under exclusive licence to Springer Nature Switzerland AG 2020
W. Truszkowski et al., *Robot Memetics*,
SpringerBriefs in Electrical and Computer Engineering,
https://doi.org/10.1007/978-3-030-37952-0_6

as the basis of knowledge exchange among robots and of knowledge transfer from humans, allowing growth in group intelligence and robot functionality.

In general, when robots gain the capability to perform memetic learning according to the requirements and principles laid out in earlier chapters, memes will enable robots to evolve their knowledge. In turn, the culture of humans and robots in a shared society will flourish.

While the broader space exploration community considers the multiple facets of how this flourishing society can be established, this exposition offers perspectives focused on its envisioned intelligent robotic members and how to realize the higher levels of individual cognition and collective intelligence that they would need to work together and with humans. It should be expected that the associated concepts and ideas about robot memetics would apply to applications and environments other than those involved in the planetary surface exploration perspective provided herein. With sufficient focused development, robot memetics will apply to many situations involving extended human and robot coexistence in common environments and close proximity.

Further Reading

1. NASA (July 2015) NASA Technology Roadmaps: TA 4: Robotics and Autonomous Systems. Retrieved from https://www.nasa.gov/sites/default/files/atoms/files/2015_nasa_technology_roadmaps_ta_4_robotics_and_autonomous_systems_final.pdf.
2. Aunger R (2002) The electric meme: a new theory of how we think. Simon and Schuster
3. Blackmore S (2000) The meme machine. Oxford University Press, USA
4. Dawkins R (1989) The selfish gene. Oxford University Press
5. Gabora L (2002) The beer can theory of creativity, in creative evolutionary systems, edited by Peter Bentley and David Corne. Morgan Kaufmann
6. Gunders J, Brown D (2010) The complete idiot's guide to memes. ALPHA
7. Hougen DF, Carmer J, Woehrer M (2003) Memetic learning: a novel learning method for multi-robot systems. Robotic Intelligence and Machine Learning Laboratory, School of Computer Science, University of Oklahoma, Norman, OK
8. Silby B (2000) What is a meme? Retrieved from http://www.def-logic.com/articles/what_is_a_meme.html
9. Stuckenshmidt H (2005) Information sharing on the semantic web. Springer
10. Truszkowski W, Rouff C, Hasan M (2014) Memetic engineering as a basis for learning in robotic communities. In: Proceedings of AIAA SCI Tech 2014. 13–16 Jan 2014
11. Truszkowski W, Hallock H, Rouff C, Karlin J, Rash J, Hinchey M, Sterritt R (2009) Autonomous and autonomic systems: with applications to NASA intelligent spacecraft operations and exploration systems. Springer
12. Truszkowski W, Hinchey M, Rash J, Rouff C (2006) Autonomous and autonomic systems: a paradigm for future space exploration missions. IEEE Trans Syst, Man, Cybern, Part C 36(3):279–291
13. Vassev E, Hinchey M (2014) Autonomy requirements engineering for space systems. Springer
14. Wilson E, Unruh W (2011) The art of memetics. lulu.com
15. E. Tunstel, J. M. Dolan, T. Fong, and D. Schreckenghost. Mobile robotic surveying performance for planetary surface site characterization. In Proceedings of the 8th Workshop on Performance Metrics for Intelligent Systems, pages 200–205, New York, NY, 2008. ACM.
16. Schenker PS, Huntsberger TL, Pirjanian P, Baumgartner ET, Tunstel E (2003) Planetary rover developments supporting Mars exploration, sample return and future human-robotic colonization. Auton Robot 14:103–126
17. Akbarzadeh-T M-R, Kumbla K, Tunstel E, Jamshidi M (2000) Soft computing for autonomous robotic systems. Comput Electr Eng 26(1):5–32
18. Tunstel E, Maimone M, Trebi-Ollennu A, Yen J, Petras R, Willson R (2005) Mars exploration rover mobility and robotic arm operational performance. In: Proceedings of the IEEE International Conference on Systems, Man, and Cybernetics, Waikoloa, HI, Oct 2005, pp 1807–1814

© The Author(s), under exclusive licence to Springer Nature Switzerland AG 2020   71
W. Truszkowski et al., *Robot Memetics*,
SpringerBriefs in Electrical and Computer Engineering,
https://doi.org/10.1007/978-3-030-37952-0

19. Maimone M, Biesadecki J, Tunstel E, Cheng Y, Leger C (2006) Surface navigation and mobility intelligence on the Mars Exploration Rovers. In: Intelligence for space robotics. TSI Press, Albuquerque, NM
20. Shifman L (2014) Memes in digital culture. MIT Press, Essential Knowledge Series
21. Stanovich KE (2004) The Robot's rebellion. The University of Chicago Press
22. Kephart JO, Chess DM (2003) The vision of autonomic computing. Computer 36:41–52
23. Samani H, Saadatian E, Pang N, Polydorou D, Fernando ONN, Nakatsu R, Koh JTKV (2013) Cultural robotics: the culture of robotics and robotics in culture. Int J Adv Robot Systems 10:400
24. Grenander U (2012) A calculus of ideas: a mathematical study of human thought. World Scientific
25. Kennedy J, Eberhart R (1995) Particle swarm optimization. In: Proceedings of the Fourth IEEE International Conference on Neural Networks, Perth, Australia. IEEE Service Center, pp 1942–1948
26. Kennedy J, Eberhart RC, Shi Y (2001) Swarm intelligence. Morgan Kaufmann Publishers, San Francisco
27. Kilroy was here. Wikipedia. Download March, 2017. URL https://en.wikipedia.org/wiki/Kilroy_was_here
28. Alex Mercer. Business Cat. http://knowyourmeme.com/memes/business-cat.
29. Golbeck Jennifer (2008) Trust on the World Wide Web: a survey. Found Trends Web Sci 1(2):131–197. https://doi.org/10.1561/1800000006
30. Vanderelst D, Winfield AFT (2016) Rational imitation for robots. In: Tuci E, Giagkos A, Wilson M, Hallam J (eds) From animals to animats 14. SAB 2016. Lecture notes in computer science, vol 9825. Springer, Cham. https://link.springer.com/chapter/10.1007/978-3-319-43488-9_6
31. Winfield AFT, Erabs MD (2011) On embodied memetic evolution and the emergence of behavioral traditions in robots. Memetic Comp 3:261–270. https://doi.org/10.1007/s12293-011-0063-x
32. Feng L, Ong Y-S, Tan A-H, Chen X-S (2011) Towards human-like social multi-agents with memetic automaton. In: 2011 IEEE congress of evolutionary computation (CEC), pp 1092–1099.
33. Hou Y, Feng L, Ong Y-S (2016) Creating human-like non-player game characters using a memetic multi-agent system. In: 2016 International Joint Conference on Neural Networks (IJCNN), Vancouver, BC, Canada, pp 177–184
34. Feng Liang, Ong Yew-Soon, Tan Ah-Hwee, Tsang Ivor W (2015) Memes as building blocks: a case study on evolutionary optimization + transfer learning for routing problems. Memetic Comp 7:159
35. Bringsjord S, Licato J, Govindarajulu NS, Ghosh R, Sen A (2015) Real robots that pass human tests of self-consciousness. In 2015 24th IEEE International Symposium on Robot and Human Interactive Communication (RO-MAN), Kobe, Japan

Printed in the United States
By Bookmasters